Date Due

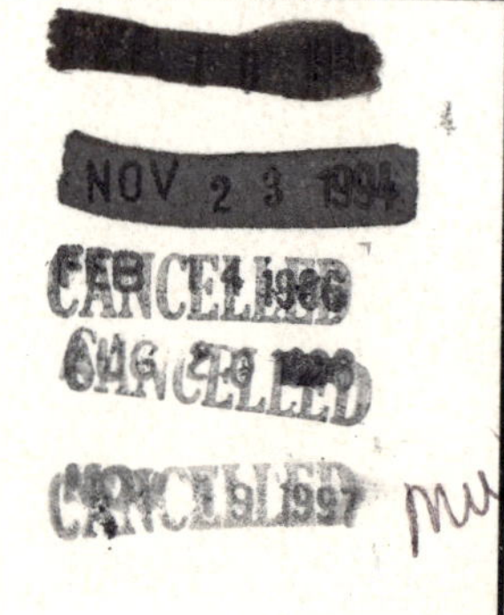

NOV 2 3 1994
CANCELLED
CANCELLED
CANCELLED

CANCELL
JAN 6 1998
CANCELL
CANCELLED
1998

DIGITAL CONTROL DEVICES

EQUIPMENT AND APPLICATIONS

J. A. MOORE

ISA PRESS

DIGITAL CONTROL DEVICES
Equipment and Applications

INSTRUMENT SOCIETY OF AMERICA
67 Alexander Drive
P.O. Box 12277
Research Triangle Park, NC 27709

Library of Congress Cataloging-in-Publication Data

Moore, J. A.
  Digital control devices.

  Includes bibliographies.
  1. Digital control systems.
  2. Process control.   I. Title.
TJ223.M53M67   1986   629.8'95   86-10542
ISBN 0-87664-901-0

The jacket design is based on a photograph from the January 1986 cover of *InTech*, which was provided by DAS, Inc., Boston, MA.

Book design by Summit Technical Associates, Inc.

FOREWORD

This text is intended as an introduction to digital control devices for people not intimately familiar with the details of digital equipment. The original concept was to describe and explain the devices that perform digital control functions, with some applications to exemplify the use of the equipment. The material in Units 6 through 11 was written to provide this, with a few examples of application.

It became apparent very quickly that there were constant references in the digital and communication context that the readers probably would not be familiar with. Units 2 through 5 were written to fill that need. Finally, there are so many things that can be done by an integrated system of digital devices that it became important to say so, and the application section lengthened from one short unit to two longer ones.

The Exercises, also, developed a function that was not originally intended. There must be a certain degree of abbreviated treatment in a text that is trying to cover a lot of ground without going into tremendous detail. The Exercises have been used to introduce some of that detail without interfering with the flow of the text. An instructor using the text should not expect his students to be able to answer all of the questions; but the students are encouraged to look at the answers for some further insight into the subjects they are reading about.

I am grateful to many people for the help I received in assembling the material for this text. The natural tendency of most people is to want to help — a few say that they will, then don't, but the majority go out of their way.

In particular I want to thank Russell W. Wolfe, R & D Systems Design, Inc., for the time spent in making an objective review of the original manuscript. His advice and comments were invaluable in establishing important contexts and avoiding misconceptions.

My very special thanks to Frank DeMarco, of Allen-Bradley; Jerry Melcher, of Acurex Corporation; Brett Holcomb, of Nekoosa Papers, Incorporated; and to Mike Ewald and Bill Buzzard, of Fischer and Porter Company. Thanks also to Ralph Hartman, Larry Heine, Wynn Boetcher, and Ed Otto, of Leeds and Northrup. I want to thank the ISA Publications Department for their encouragement and for making much valuable reference material available to me; and Pat Kaltwasser, my local librarian, for chasing reference books half way across the country for me — and finding one in Marcus Hook, PA (shades of E. B. White).

J. A. Moore
January, 1986

Contents

Tables

Figures

UNIT 1
Introduction

1-1 The Increasing Importance of Digital Instrumentation and Control

Trade journals provide an accurate indication of the state of their field of interest at the time of publication by reporting the latest developments in that field. The September, 1970, issue of the instrumentation magazine *Instruments and Control Systems*, for example, presented its readers with articles about Analog Computers in Process Applications, Temperature Measurements by Infrared Radiation, FET Conductance Measurements, Cam Shape Synthesis, Precision Timing Systems, Digital Position Measurements, and a Review of Program Controllers. There was a preview of the upcoming ISA International Conference and Exhibit. Notices of new products included electronic pressure transmitters, butterfly valve actuators (3–5 psi), CRT display drivers, teleprinters, and integrated circuits. There were a number of advertisements for pneumatic tubing and fittings, some for filled system transmitters, and a number for indicating controllers. Digital equipment was not left out; other advertisements extolled the virtues of tape readers, 24-bit computers, integrated circuits, and data acquisition systems using indicators with LED readout. In summary, digital equipment was in use, although more for laboratory and aerospace applications than for process control applications.

In contrast, the June, 1985, issue of *I and CS* (the Industrial and Process Control magazine) featured a Temperature Monitoring and Control Update, and articles about networking (The Higher Level Network Layers), computer graphics (Buying the Resolution You Need), bus boards (What's New in Bus Boards?), and data loggers (Data Logger Helps Reduce Costs in Brewery). It carried advertisements for 3 programmable logic controllers, 5 printed circuit cards and digital hardware devices, 3 distributed control systems, 6 digital PID controllers, 1 data acquisition system, 1 strip chart recorder with digital circuitry, 4 devices supporting personal computers, 1 programming language, and 2 digital-to-analog converters. New products described included 7 programmable logic controllers and support equipment, 1 distributed control system, 16 printed circuit cards and digital hardware devices, 1 single-loop digital PID, 3 data acquisition systems, 13 devices supporting personal computers, 3 communication support devices, a process management interface to distributed control systems, and a personal computer. Filled system self-contained temperature regulators, and pneumatic tubing were still advertised.

The emphasis in 15 years has swung decidedly from analog to digital implementation of instruments and control. Everyone working in the field has been conscious of the change, but only a comparison with a relatively recent past demonstrates the degree to which it has taken place. Not very long ago a potential instrumentation technician or engineer learned about flapper/nozzle systems so that he could properly apply pneumatic equipment. He studied feedback control theory, read up on frequency response, Nyquist plots, and root locus, and became conversant with the subjects of time constants and response times of components and processes. He learned about transistors and operational amplifiers and was able to size control valves and orifice plates. He could construct ladder diagrams and then learned how to design the equivalent circuits using Boolean algebra. Today, a working knowledge of digital computers has been added to the required reading — and understanding — that makes up the list of technologies and tools that this discipline demands.

This next step in learning is not a matter of choice — digital computers have moved into our lives. This is the context in which high school graduates are already thinking — and digital control devices are the type of equipment that plant management is buying today.

A lot of new equipment has to be bought "right now," because the older instruments are wearing out and must be replaced. Many systems of pneumatic control that have performed faithfully for twenty or thirty years — and satisfactorily, in the context of past needs — can no longer be operated reliably because spare parts have been used up, and replacements are no longer being manufactured. New equipment must also be bought right now for plant facilities that are being expanded, where new processing installations are being built. In either case, retrofit or new plant, there is a very good chance that the new equipment will be microprocessor-based. The need to become familiar with the way it functions is immediate.

An indication of the importance of digital control devices and techniques in instrumentation is evidenced by the number of Short Courses already available to students interested in adding an understanding of digital instrumentation to their store of knowledge. Table 1-1 lists courses included in the ISA Training Programs for 1984–1985.

As a contribution to the general subject for which the courses listed present specific details, this book is intended to put into perspective the categories of equipment that perform digital control for the process industries, their functions, and their applications.

Table 1-1
ISA Short Courses Related to Microprocessor-Based Instrumentation

S450	Introduction to Distributed Process Control Systems
S600	Basic Elements of Digital Systems
S620/T6200	Principles and Applications of Digital Devices
S630/T6300	Microprocessor Fundamentals
S640	Microprocessors in Control Systems
T6500	Microcomputer Programming for Process Control
T6600	Digital Control Systems
S670/T6700	Industrial Computer Control Strategies
S680/T6800	Introduction to Programmable Controllers
T6900	Introduction to Robotics

1-2 Reasons for the Acceptance of Digital Instrumentation

Analog control equipment will not be completely replaced by digital control devices for a long time. At the present time process variables are measured, and the signals transmitted, almost entirely by analog devices. Regulation is performed entirely by analog regulators. For some very fast process requirements, such as compressor surge control, real time analog control is always recommended. For small systems, analog control installations are less expensive than digital ones. Nevertheless, digital control devices have characteristics that are making their use the dominant choice for future installations. There are a number of reasons for this:

(1) Flexibility provided by programming.

A. Flexibility in the application of digital equipment.

Cutting a panel opening to change its shape takes time and effort. So does changing wiring connections to get a different result from an instrument action. On the other hand, giving something a new name is an easy thing to do. This is the power of software. Flexibility provided by programming is an important attribute of digital equipment. Systems as built and installed rarely perform exactly as they were designed to, even when their design repeats that of an apparently identical unit. There are always some functions that were forgotten, or misunderstood, or could be changed to the advantage of production and profit. This is costly and time consuming in a hard-wired analog system. Holes have to be cut or filled up in panels, work orders have to be written and purchase orders issued, and often changes can only be made during an outage, when equipment is not operating. A digital system allows many changes to be made from the keyboard terminal while equipment is on line.

B. Flexibility in measurement of process parameters.

Because they can be programmed, digital control devices can use measured variables to compute other variables that represent the true conditions of a process but are unmeasurable because transmitters do not exist that can make the measurements. Use of these computed values can increase productivity because fewer control loops will have to be dependent upon inferential measurements.

C. Flexibility in the performance of control.

Analog control, by itself, functions poorly when there are long transport lags in a system and when variables in a system interact. Many installations of analog control, applied to processes with these characteristics with the best designed intentions in the world, operate with the instruments in the manual control mode because the operators can perform better than the instruments can. An experienced operator, instinctively able to anticipate changes in load and demand that his process undergoes, can optimize control by combining his knowledge with his observations and make changes manually that produce satisfactory results. He is doing what a computer can be programmed to do but what a group of individual analog loops cannot do.

There is an obvious limit to this application of expertise. A good operator may be able to produce cement or glass of satisfactory quality in this fashion, but anything faster and more complex is certainly beyond human capabilities. However, with minicomputers economically justifiable as a solid state electronics continues to reduce pricing, optimizing with digital computers used as supervisory controllers can put these control systems back on automatic control.

(2) An improved man-machine interface.

An operator depends on the information he sees on his instruments. Digital equipment avoids some recognized deficiencies in the way analog instruments present information to an operator.

Analog instruments are read by interpolating the position of a pointer between divisions on a scale. Digital readout is accomplished by actuating light emitting diodes to create digits that have an absolute meaning. Furthermore, these numbers do not change because of drift, as may happen with analog value readout. Admittedly, the analog signals from which these numbers are derived may be in error. From the operator's point of view, however, there is a decrease in uncertainty, and an increase in confidence, and this will usually result in increased productivity.

(3) Improved controllability.

Digital instruments have performance characteristics that can improve product quality over the level attained when analog equipment is used. Where the differences between digital control and analog control equipment offer advantages, instrument manufacturers have been able to use them in the design of their products to improve the quality of automatic control.

Analog circuits operate in real time; the signal measured continuously represents what is occurring at the instant of measurement. This is an advantage of analog equipment. But analog circuit components (resistors, transistors) and soldered connections change their characteristics with temperature, and with age. Therefore, signal values drift. Solid state circuits do not drift to the same degree, if at all. Analog values must be set by mechanical adjustments and read as pointer values interpolated between scale divisions. Digital displays are precise. Digital values change instantaneously, as binary numbers change; analog value changes have a finite time delay because they must pass through all the values from the original to the final value when there is a change. Therefore, analog circuits must operate more slowly than digital circuits.

Many analog devices cannot accept low level signals, Therefore, the outputs of low level transmitters must be amplified and, in most cases, linearized. This disadvantage puts additional components in the measurement chain, and more error.

The ability to indicate a precise digital value allows higher resolution to be used in setting reference control point values. Values converted from analog process variable measurements can be made very precise to match these set points. Digital values displayed will differ, at most, from values measured by a change in the value of the last significant bit of a digital word. 12-bit words are commonly used by analog-to-digital converters, so accuracy is one part in 4000. This will become increasing more precise as digital transmitters are developed.

These operating parameter advantages of digital over analog instrumentation add up to a reduction in uncertainty. Users who have been clamoring for readability of ± 1 degree in a range of 0 to 2000 can get that sort of resolution now. Some are finding that they may not have process equipment that can take advantage of such small changes in reading.

(4) The capacity to store data.

Digital control devices are particularly beneficial for applications that require many variations of the same manufacturing procedure (batch control using recipes), documentation of process parameters formatted to fulfill specific management needs, the collection of large amounts of data, and performance of computations that enhance supervisory control. Batch processes, which represent a large proportion of the industrial processes that use automatic control equipment, need one or all of these attributes.

(5) The power of expressing numbers digitally.

An arrangement of 1's and 0's can represent many things, depending on the context in which it is used. A byte (eight consecutive bits) has 256 possible definitions. It can represent 256 combinations of discrete input signals; 256 addresses; 256 commands; 256 pointers to sequences of information movement. It can be argued that a decimal number is also powerful; it can refer to a page of a book or, indeed, to an entire library. The number by itself, however, has only one meaning as a simple integer value and is much more limited.

(6) The facility with which binary data can be moved about.

Data in digital systems is expressed in binary numbers. Binary numbers, having only two values, can be easily represented by two values of energy. Techniques exist for moving large amounts of information in the form of electrical or light energy at high rates of transfer, and over long distances. Because of this, using digital instrumentation extends the scope of control from a local process area to the boundaries of entire industrial installations.

(7) Improved troubleshooting and maintenance techniques.

Digital circuits can be programmed to perform self-diagnostics as a background activity. This provides them the capability to continually check themselves, monitoring for errors and faults. They can announce their problems; display messages that identify the source of the problem; and in some cases switch automatically to alternate paths of operation or to alternate spare components so that process operation is not interrupted or degraded by component failures.

Maintenance and troubleshooting procedures for digital systems are similar to those for analog systems, in that a problem is found by isolating and testing. Because of the complicated design and structure of digital equipment, however, it has become customary to troubleshoot to the point where a fault can be localized to a single circuit board and the problem corrected by making a circuit board replacement. This reduces down time, involves a small spare parts inventory, and improves system reliability.

(8) Reduced manufacturing costs.

Manufacturing techniques that allow large amounts of circuitry to be compressed onto a minute silicon area have allowed computers to be mass produced and have reduced the size of the assembled product. Both mass production and reduced size strongly promote cost reduction. The small size, in addition, results in reduced power dissipation, which in turn means lower operating temperatures. This increases the life of components and improves operating characteristics.

(9) Reduced installation costs.

Since digital circuits can be built on silicon chips while analog devices must be mounted and wired together, the digital control system can be packaged in a small housing. In contrast, the analog system components must be mounted in panel assemblies and interconnected by expensive wiring. The communication networks that interconnect digital components over long distances can consist of a data highway made of a single coaxial cable or a single fiber optic cable. In contrast, each analog device must have at least two wires to bring its signal information to another device.

1-3 Defining Digital Control

What does "digital control" mean in this text? First, though, what does it *not* mean?

Look through the reference cards in a technical library file and you will find that the term "digital control" usually means one of two things. Used in the context of continuous control of process variables, it refers to the **mathematical theory** of dynamic control using control functions programmed in software. Used in the context of batch, sequential, or discontinuous control, it refers to the **design** of control systems made up of discrete functions, functions that occur in one of two possible states. This book is *not* specific to "digital control" in either of these contexts.

The first of these two meanings of digital control comes about because control functions programmed for digital computer implementation respond to process changes in ways different from control functions performed by analog devices. Inputs and outputs to digital controllers are discontinuous, while the processes to which they apply exist in real time, varying over a continuous range of value. Inputs to digital computers, instead of being continuously read as would be the case for an analog system, are sampled. Outputs, instead of being continuously applied, are updated and held. This on-off characteristic of digital computers causes them to respond and perform differently from analog devices used in real-time control of dynamic systems.

Books written to explain the advantages and disadvantages of using digital devices for feedback control, employing the techniques required to achieve satisfactory dynamic response and minimum error in a sampled data system, are intended for instruction of the serious student and for guidance for the control system designer. Some go so far as to place considerable emphasis on the z-transform, require some understandng of Laplace transforms, delve into the nonlinear phenomenon of amplitude quantization and its effects on system error and dynamic response, and put discussions of eigenvectors and the Cayley-Hamilton theorem in appendices for the edification of the serious student. This is more advanced material than will be found in this text.

The second use of the term "digital control" exemplifies design criteria for control systems that use discrete signals — inputs and outputs that are either on or off. The activity in these systems usually occurs as a sequence of operations that involve counting, timing, and the opening and closing of contacts to pass or block power flow. Digital control in this context describes the design of logic circuits, the combination of on-off conditions, to accomplish desired control actions. Books dealing with this form of digital control may discuss Boolean algebra, truth tables, and circuit optimization using Karnaugh maps.

Many good books are available to those interested in these two aspects of digital logic control. This book, however, has different objectives. It is written in the context of using, not designing, instruments. Its subject is the instrumentation used for control of industrial processes, instrumentation that has been performed traditionally by analog devices but which more and more will be designed in accordance with the technology represented by the digital control devices referred to above. Accordingly, this book is about control using microprocessor-based equipment — the specific categories of digital devices used, how they function, and how, while appearing to perform the same applications as analog equipment, they have hidden talents to accomplish many activities that analog devices cannot.

All of the microprocessor-based devices performing instrumentation functions are really special-purpose computers. In this text, they will be referred to as "digital control devices".

1-4 Information Flow

The text is principally concerned with describing the form and applications of the digital control devices that have evolved in their present state because of the designs made possible by using microprocessors. This is not the whole story, by any means.

A second concern of the text is the dependence of digital control on information flow. Digital control is a manifestation of a dual revolution, one aspect of which is the integrated circuit, the silicon chip. The other side of the coin is the network of communications that interfaces digital devices, combining many separate pieces of hardware into one integrated system that can move unbelievably large amounts of information in an unbelievably short time. These two revolutions are taking place simultaneously and changing our entire social structure as they do. The microprocessor has changed the forms of devices, and communication developments have changed the ways they are used.

The microprocessor, heart and brain of the computer body, exists to move information, to make it flow from one point in space and time to another. The same statement is true for all computers, logically, since the microprocessor is a computer in miniature. Digital control devices are special-purpose computers, and so the statement extends to them.

In all systems where digital control devices are used for process control, information about the process conditions comes from sensors at the process. Discrete on-off signals come as digital information, and analog values are converted to digital form. The special-purpose computer processes the digital information, stores it, and develops digital values for outputting to peripheral devices and for performing process regulation. The outputs designed for regulating the process are changed back to analog values. This is the familiar feedback control loop in another form; the implementation has changed, but the objectives are the same.

Information flow, as illustrated in Figure 1-1 — between memory and processors, between instruments making up a system — becomes a basis for talking about the common characteristics of the different categories of devices that will be examined in the text.

1-5 Text Content

This book intends to introduce to the potential user the digital control devices that are replacing analog control devices for instrumentation and control — what they are and how they are being used. The text is introductory, directed toward the person who has had limited exposure to digital control applications, or who is about to become exposed to them. In addition to its interest to students, portraying a part of the overall perspective of instrumentation and control, it is intended be useful to an instrument technician or control system designer who has experience in conventional analog instrumentation but little or none in digital implementation; a project manager about to be assigned for the first time to a project where digital control equipment will be used; or a purchasing agent who is going to have to evaluate bids for digital control equipment. An objective of the text is to prepare such individuals so that they become conversant with the subject to the extent that they can make objective comparisons and decisions. It is not intended to make them expert designers of digital equipment, and it will not tell them how to troubleshoot digital circuits.

The subject matter of the book, like all Gaul, is divided into three parts. There is a semi-technical section, expository for digital techniques; a section specific to digital control devices; and a section devoted to applications.

1-6 Digital Techniques

Books designed for advanced studies can assume that the reader is thoroughly grounded in the fundamentals of the subject about which they are written. Those written specifically for a less experienced group of readers must include some background material, which can be skipped or skimmed if it is already part of the reader's store of experience.

Each of the background material subjects listed in Table 1-2 is covered in a separate chapter. Each chapter is common to the description of the specific categories of digital control devices discussed in the second portion of the text and is intended to create a context to which all the following discussions can be referred. The subjects presented in these preliminary chapters are covered not in detail, but only to the extent that information movement can be emphasized and exemplified.

Table 1-2
Technical Aspects of Digital Control Devices

Unit No.	Subject
2	Binary Notation
3	Hardware and Software
4	Systems; Single Instruments and Control Systems
5	Communication

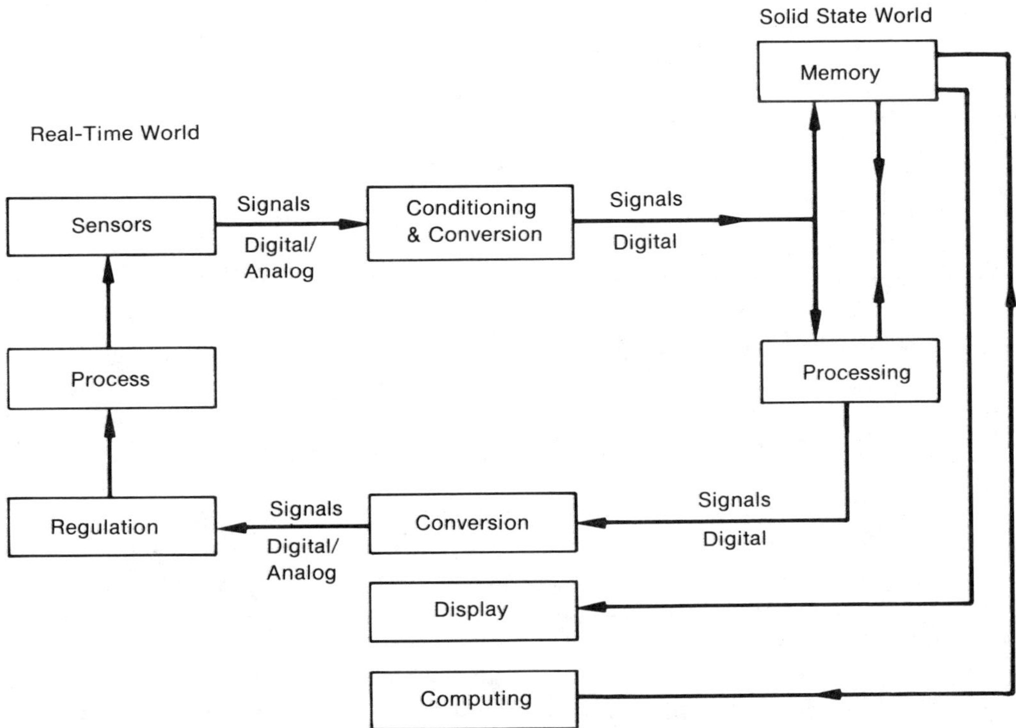

Figure 1-1. Information Flow through a Digital Control System

Binary Notation

When the concepts of counting in binary and of storing information as a combination of 1's and 0's are understood, the usefulness of the binary notation format becomes apparent. The discussion of this section provides a basis for that understanding.

Hardware and Software

In developing subsequent sections about the different products that qualify as digital control devices, it became apparent that they all were made up of subcomponents that performed similar operations: input/output processing, analog-to-digital conversion, and microprocessor-based central operating units. Instead of explaining what the parts were each time the terms were encountered, it seemed easier to do it once in a general section about hardware.

The discussion begins with switches, the most basic of digital functions. The basic switching device is combined into larger subcomponents for decision making and for memory retention. Progressively, larger components (assemblies that process information) are discussed.

An introduction to the subject of software is included in this section as an extension to the theme of information flow.

Systems

In the context of process control, components by themselves accomplish very little. Only when they are incorporated into a system where there are measurements to provide information and other devices to receive the outputs their operation has developed do they contribute to the making of a product. Application has to be considered in the context of the system in which it is used, and so it is important to explain what a system is. Accordingly, a section has been included that first discusses the hardware system created by combining appropriate subsystems discussed in the preceding section into an assembly (the system that we have defined as a "digital control device") that performs a specific purpose. The larger system, which uses the digital control device for industrial process control by combining the specific component and its peripheral equipment with an industrial process as well as with other devices to provide inputs and outputs, is then addressed.

Communication

Continuation of the discussion of information movement between devices explains how information is gotten out of a device, changed into a format that can be sent over interconnecting cables, sent over the interconnections in a fashion that gets it to the proper destination, verifies its correctness, and makes sure that all devices in a system can communicate as they are required to. Communication over networks can be a confusing subject, not because it is unstandardized but because it is strictured by a great number of standards. The objective of this section has been not to explain the many standards and how to use them, but to show the purposes that they fulfill and to make the terms they use recognizable to the reader.

1-7 Digital Control Devices

The primary objective of the text is to examine the types and applications of microprocessor-based equipment that are being used to collect process information and control industrial processes, both batch and continuous. Categories of devices discussed, in increasing order of complexity, are listed in Table 1-3.

In order to establish frames of reference, it has been necessary in some instances to include specifications as well as examples of the application of specific equipment. In describing what a particular class of equipment will do, it has been necessary to include some characteristics that may be specific to one supplier in order to make the illustration as

Table 1-3
Categories of Digital Control Devices

Unit No.	Category	Function	
		Primary	Secondary
6	Data Acquisition	Data Gathering	Data Processing
7	Programmable Control	Relay Logic	Process Control Computation
8	Single-Loop Control	Proportional Control	Advanced Control Computation
9	Distributed Control	Continuous Process Control	Sequential Batch Control
10	Minicomputers	Optimizing, Report Formatting, Modeling, Supervisory Control	Management and Scheduling
11	Personal Computers	Network Coordination, Data Processing	Spreadsheets, Word Processing

realistic as possible. Every effort has been made to keep such information in general terms and to make value and descriptions typical, not specific. No supplier is endorsed by the author or by ISA, and any similarity to actual products is coincidental, unless there is no other way to make a point.

This portion of the text, which talks about the devices that implement digital control systems, includes descriptions of the process functions the devices perform. For example, it may be assumed that the reader who is not familiar with feedback control theory will not get very much out of the statement that a device is used for PID control if he does not understand what PID control is and how it is used in process control. Accordingly, the section describing each device that accomplishes specific applications includes some description of the application as background material. Again, the reader with experience will skip this explanation.

1-8 Applications

Chapters presenting applications for which digital control devices have been used complete the text. The applications have been selected to illustrate those characteristics of digital control instrumentation that are making it uniquely useful for process control. In other words, they are taken from processes that require much data gathering, numerous recipe changes, special formatting of reports, coordination between different branches of plant operation, and a significant amount of mathematical computation.

It is not possible to match industries with categories of digital control devices in such a way that there is any one-for-one correspondence. It will be evident that more than one category of digital control device can be used to perform each task used as an example. It will also be evident that there are a number of levels of sophistication that make minicomputers most practical for certain classes of application and programmable logic controllers more practical for others. To that extent, there is a correspondence in the sense that more sophisticated needs are filled by more sophisticated intrumentation.

The application examples in Unit 12 are equipment oriented; that is, they are descriptions of uses to which specific categories of digital control devices have been put. Applications are included for using data acquisition units, single-loop controllers, programmable logic controllers, and distributed control.

The illustrations used in Unit 13 address the application of digital control equipment to the needs of complex processes and interrelated processing areas. Waste treatment, pulp and paper, and energy management are representative of this degree of complexity. An example of a heat-treating installation makes use of several categories of digital control equipment coordinated by a personal computer. The section concludes with a brief review of hierarchies of control and an anticipation of its extension to fully automated plants.

REFERENCES

1. Liptak, Bela, Editor, *Instrument Engineers' Handbook*, Chilton Book Company (1985).
2. Balch, Thomas, Andreas Schreyer, and Jerry Tonn, "The Impact of Semiconductors on Industrial and Process Control," *Instruments and Control Systems*, September, 1981, pp. 80–88.

EXERCISES

1. List variables that are important to process control but cannot be readily measured. Show how they can be computed from other variables for which measurements can be made by commercially available devices.

2. Why is the drift-free characteristic of the digital PID an important characteristic for closed-loop control?

3. How would you go about finding the break-even point in the number of loops of analog vs. digital control, on the basis of cost?

4. What are the constraints and considerations when retrofitting, replacing a pneumatic control system with digital control devices?

5. Using published information from vendors' catalog material or instruction manuals, compare the specifications for an analog proportional, integral, derivative (PID) controller and for the digital equivalent.

UNIT 2
Digital Notation

Using only the digits 1 and 0, any quantity can be represented as a number. 1 can also represent the state of being, and 0 the state of not being. Digital computer design uses these complementary characteristics by putting together combinations of electrical energy that can be read as numbers, but also used as codes to activate circuits.

It is not necessary to be able to use binary numbers for counting in order to understand how digital equipment is applied in industrial process control; insisting on a detailed understanding would certainly obscure the forest by creating too many trees. However, in subsequent sections of this text, terms will be used that refer to these coded operations, the sizes and configuration of the combinations, and the ways that they are manipulated. If there are readers who have not used digital numbers very often the terms may not be completely clear, and the continuity of the description could be lost. To obviate this possibility a summary of digital notation has been included as background, to provide a perspective for understanding the use of digital numbers for expressing ideas as well as numerical values.

2-1 Numbers in Decimal Notation

Starting with familiar structures, decimal notation is based on powers of 10. There are ten digits, 0, 1, 2, 3, 4, 5, 6, 7, 8, and 9. The lowest order digit of a decimal number is multiplied by 10 raised to the zero power, the next lowest is multiplied by 10 raised to the first power, and so on. The number 527, for example, has a value made up as shown in Figure 2-1.

2-2 Numbers in Binary Notation

In exactly the same way, binary notation is based on powers of 2. There are two digits, 0 and 1. The lowest order digit of a binary number is multiplied by 2 raised to the zero power, the next lowest is multiplied by 2 raised to the first power, and so on. The number 1000101, for example, has a decimal value made up as shown in Figure 2-2.

```
527

     7*10⁰  =  7*1    =    7
     2*10¹  =  2*10   =   20
     5*10²  =  5*100  =  500
                         ----
                          527
```

A decimal number is formed by combining powers of 10.

Figure 2-1. Decimal Notation

```
1000101

     1*2⁰  =  1*1    =   1
     0*2¹  =  0*2    =   0
     1*2²  =  1*4    =   4
     0*2³  =  0*8    =   0
     0*2⁴  =  0*16   =   0
     0*2⁵  =  0*32   =   0
     1*2⁶  =  1*64   =  64
                         ---
                          69
```

A binary number is formed by combining powers of 2.

Figure 2-2. Binary Notation

In Figure 2-2 a binary number is converted to a decimal number. Figure 2-3 illustrates decimal-to-binary conversion.

Since numbers can be negative as well as positive, there must be a way to tell them apart. To differentiate, a binary number may be coded, using the left-most digit. 1 is used for (−) and 0 for (+). Eight digits in a coded combination can represent a number from −128 to +127, as shown in Figure 2-4.

This form of notation is called "twos complement," so named because the negative numbers are written in a manner that complements the positive numbers. To get the twos complement of a 7-digit binary number, change each 0 to 1, change each 1 to 0, and add 1. The twos complement of +5, 00000101, is 11111010 + 1 , 11111011 or (−) 5.

2-3 Bits and Bytes

Each individual 1 or 0 may be called a bit, and binary values are often stored and transferred eight bits at a time. A group of eight bits is called a byte. Half a byte is sometimes

```
Starting with the decimal number 37,
           37 divided by 2  =  18 +      1
           18 divided by 2  =   9 +      0
            9 divided by 2  =   4 +      1
            4 divided by 2  =   2 +      0
            2 divided by 2  =   1 +      0
            1 divided by 2  =   0 +      1
      Written as                    100101
```

Figure 2-3. Converting Decimal Numbers to Binary Numbers

0	1	1	1	1	1	1	1	+127
			.					
			.					
			.					
0	0	0	0	0	1	0	0	+ 4
0	0	0	0	0	0	1	1	+ 3
0	0	0	0	0	0	1	0	+ 2
0	0	0	0	0	0	0	1	+ 1
0	0	0	0	0	0	0	0	+ 0
1	1	1	1	1	1	1	1	− 1
1	1	1	1	1	1	1	0	− 2
1	1	1	1	1	1	0	1	− 3
1	1	1	1	1	1	0	0	− 4
			.					
			.					
			.					
1	0	0	0	0	0	0	0	−128

```
The left-most bit is the sign bit.
```

Figure 2-4. Twos Complement

called a nybble, although this is not common. Bits and bytes are common expressions, however. The maximum binary value that can be stored in a byte is

$$11111111$$

and this has the decimal equivalent of

$$1*128 + 1*64 + 1*32 + 1*16 + 1*8 + 1*4 + 1*2 + 1*1 = 255$$

The binary notation 100000000 is one digit larger than 11111111 and has the decimal equivalent of 256. Written as two bytes, it would be represented by 00000001 00000000.

The byte is the fundamental unit for processing information in microprocessors. Memory is usually provided in blocks of 1024 bytes. This is rounded off to 1000 and referred to as 1K of memory. A computer described in literature or advertising as a 64K machine has 64*1024 = 65536 bytes of memory.

2-4 Hexadecimal Notation

Stringing 1's and 0's together indefinitely is awkward, and the chance of making mistakes in reading or writing is great. A group of four bits is easier to handle, and another system of notation called hexadecimal has been developed using the 16 possible decimal numbers that can be represented by four bits. Figure 2-5 shows the bit groupings, hexadecimal values, and equivalent decimal values of this notation.

A binary number such as 1001001110001101 (37773 in decimal) is written as 1001 0011 1000 1101, which is 9, 3, 8, D in hexadecimal notation. 93 represents 1001 0011. It will usually be written 93H, so that it is not mistaken for a decimal number. 8D represents 1000 1101 and will usually be written 8DH. Some systems that operate on numbers in sixteen-bit groupings will write this number as 938DH, but others write the least significant byte first: 8D93H.

To convert from hexadecimal to decimal notation, remember that each hex digit has a possible value of 16, so the conversion uses powers of 16. Figure 2-6 shows the conversion of 1001001110001101, decimal 37773, written as 938DH.

Bit Grouping	Hexadecimal Value	Decimal Value
0000	0	0
0001	1	1
0010	2	2
0011	3	3
0100	4	4
0101	5	5
0110	6	6
0111	7	7
1000	8	8
1001	9	9
1010	A	10
1011	B	11
1100	C	12
1101	D	13
1110	E	14
1111	F	15

Figure 2-5. Hexadecimal Notation

2-5. Octal Notation

Hexadecimal, or hex as it is usually called, is a commonly used notation because most microprocessors operate by transferring and using 8-bit or 16-bit numbers. In another form of notation used in the days when transmission was used primarily for teletype communication, combinations of three bits are used. There are eight possible combinations of three bits, so this notation is called octal. Figure 2-7 shows these combinations with their decimal equivalents. The derivation of decimal 37773 from octal 111615 is also shown.

2-6 BCD Notation

In still another method of creating numbers, the integers that make up a decimal number are converted into binary, one digit at a time, starting with the highest power digit. The

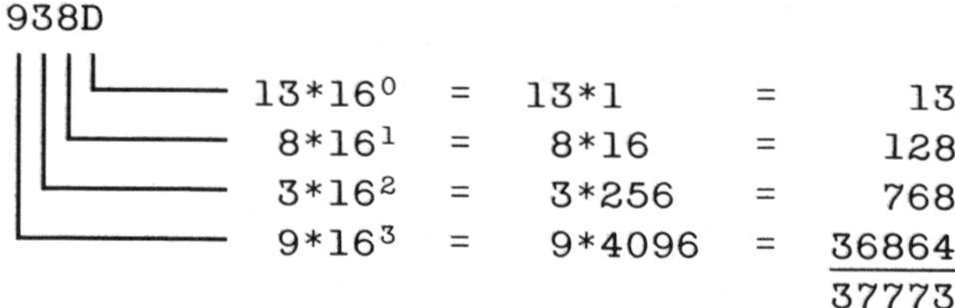

Figure 2-6. Hexadecimal-to-Decimal Conversion

Binary	Octal	Decimal
000	0	0
001	1	1
010	2	2
011	3	3
100	4	4
101	5	5
110	6	6
111	7	7

Binary 100100111000110l is combined into the groupings 001 001 001 110 001 101 and written 111615 in octal.

To convert to decimal, use powers of 8.

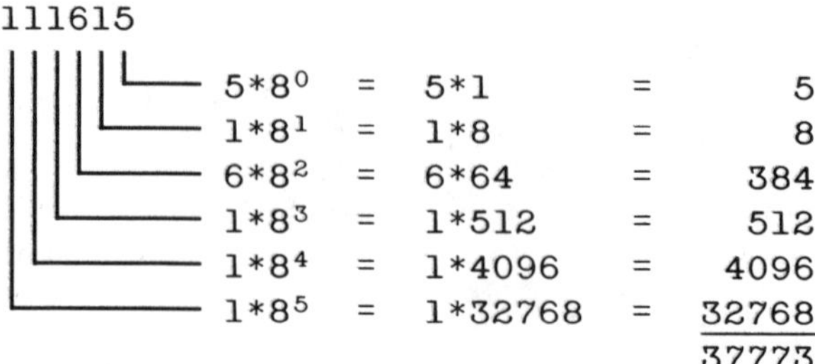

Figure 2-7. Octal-to-Decimal Conversion

formulation is called binary coded decimal (BCD). 365, the decimal representation for the number of days in a year, becomes

0011   0110   0101

in BCD. It is as though the numbers had been switched into the computer one digit at a time, and, in fact, this is exactly what happens when a thumbwheel switch is read by a digital circuit. The receiving logic is programmed to multiply each group of binary numbers by the appropriate 10's multiplier, then combine all of them into a single binary number.

2-7 ASCII Code

What information can be communicated with a byte? The numerical value is obvious. The eight bits can be arranged into 256 combinations, starting at 00000000 and continuing on to 11111111 (or if it were written in hex, 00 to FF). In addition to the numerical value, a combination can be coded to correspond to a letter, a command, or indeed, anything you want it to be. There are many codes that use hex numbers to represent letters or characters so that messages can be transmitted as a string of bits. One of the most common codes is ASCII (American Standard Code for Information), a seven-bit code that can be used to represent all the upper and lower case letters, many characters, and a number of communication functions. Data variables may be stored in computer memory in ASCII format, and, because of the inherent mathematical nature of the notation, they can be manipulated. For example, upper case "C" in ASCII is 1000011 0 (43H), while lower case "c" is 11000011 (63H). Adding 20H changes upper case into lower case. Among the characters and functions that can be represented 24H is the ASCII code for the dollar sign, "$", 0DH is used to create a carriage return, and OAH represents a line feed. The ASCII character code is listed in Appendix A.

2-8 Information Flow

The flow of information throughout a digital control device will be a recurring theme in this text. One of the great strengths of the digital design is the facility with which information can be distributed by moving combinations of binary digits from one place to another, delivering the information by virtue of the coding contained in their arrangements.

The entire activity in a computer consists of moving bytes of information from one storage location to another, adding them or comparing them, and moving the result of the manipulation to another location. Programs are lists of commands that define the movements and contain the sequence in which they shall be performed. Bit arrangements define the locations from which information shall be fetched and to which it shall go. They define the sequence of transferal. At the same time, the 1 or 0 status of specific bits in an arrangement may be used as keys to unlock the circuit hardware that is designated as the receiver, by enabling specific functions of the hardware.

How these actions are accomplished in hardware is the subject of the following section.

EXERCISES

1. How are binary numbers added?

2. What is parity, and what use is made of it?

3. If you were to encode the alphabet by assigning a different digital word to each letter starting with 1, how large a word would be required?

4. What is meant by "most significant bit" and "least significant bit"?

5. In order to start a batch sequence, the following must be true conditions:
    a. The tank agitator must be running.
    b. Pumps for three liquid ingredients must be running.
    c. There must be sufficient solid material in a bin in order to satisfy the batch requirements.
    d. The reactor tank must be empty.
    e. The drain valve must be closed.

   Seven closed switches connected to a terminal board input this information to a computer. Assuming that the terminal to which no connection is made is in the most significant position, what is the binary word that the computer will use for comparison to determine that the process is ready to start? What is the hexadecimal equivalent word?

6. A block of addresses representing the number of days in a leap year starts at address number AE20. Each address has eight bits. What address holds the number for May 19?

UNIT 3
Hardware and Software

3-1 Introduction

In a text describing lawn mowing devices — riding mowers, self-propelled mowers, and push mowers, for example — it will be evident that all of them are made from a relatively small number of basic categories of parts. Parts catalogs will show hundreds of varieties in each category, but all of the assembled devices, whether riding mowers or the walk-behind variety, are operated by some sort of motor, have blade assemblies and frameworks with handles and wheels, and can be supplied with certain attachments. Rather than defining and describing all components each time a different device is introduced as a topic, it would be proper to talk about types of components (motors, ignition systems, power take-offs, etc.) in detail in a preliminary section, then simply define specific differences in the components when the different types of mowers are described.

The same commonality of components is true for digital control devices. Later units of this text will be devoted to a number of categories of devices that perform digital control applications in distinctly different ways. As we discuss them it becomes apparent that they are all special-purpose computers. It is logical, therefore, to spend some time talking about the basic components and digital circuitry used in these types of computers, to avoid having to repeat the same subject matter for each type of device.

There are several levels of complexity in the hardware design of digital control devices.

(1) At the lowest level there are elements that perform the hardware equivalent of the TRUE or FALSE condition of the 1's and 0's. These are switches, which allow an electrical signal to pass or else interrupt it. The signals could be compared to individual letters in a written language.

(2) At the next level switches are combined into small components. These components are used to store data or to make decisions. Storage of data is called memory. Decision-making devices can be used to add or compare, the two principal activities taking place in a computer. Other decision-making components count, decode signals, and contribute to information transfer between storage areas on a timed basis.

Switching, storing, and decision-making take place in integrated circuits and are the functions performed by the circuitry built into chips. They are the building blocks of the components found in all computers. The written language equivalent to subcomponents is the word.

(3) At the highest level of circuit sophistication are assemblies built from the storage and decision-making components. They include microprocessors, multiplexers, analog-to-digital and digital-to-analog converters, and sample-and-hold components. These could be likened to sentences or even paragraphs in a written language.

Since all of these devices exist as combinations of 1's and 0's, they can be moved about in accordance with a schedule of instructions. This is called software. There are several levels of software, just as there are levels of hardware.

(1) At the lowest software level are commands that move combinations of bits to specific locations to enable specific activities (Read, or Write, for example) of components. The organization of the commands is called a program, and the language for this level of programming is called machine language.

(2) At the next level, chunks of machine language instructions corresponding to basic operations (Move a byte, or Add, for example) are defined by a coded language program. This language is called assembly language.

(3) At the highest level, languages written to express computational or functonal operations (If X then Y, or At location a,b print C) express complete actions. The computer, through internal programs, converts these commands to machine language.

These levels of hardware and software are reviewed in this section. They are supportive to the subsequent discussions of the categories of digital control equipment so that while the process relationships are being described in terms of flow of information, the flow of signals through the devices that directly perform the computer operations can be mapped onto the description by the reader.

3-2 Logical Circuits — Level 1

What are the digital circuit building blocks?

The basic function performed is switching. After all, what could be more symbolic of digital logic than a switch? It is either closed or it is not closed. Switches open and close to let current pass from high voltage sources to lower voltage levels. All the circuits that perform digital logic are made up of devices that behave exactly like switches.

Digital circuits perform logical operations. This means changing or using logic levels, a term that is measurable as a voltage but really means the condition that the voltage level represents. For digital logic, there are only two conditions. One may be represented by a high or a low voltage (the choice has to be defined; there is negative logic and also positive logic), but by definition it means that a condition is TRUE, or SET, or HI, or ON. The words all mean the same thing. The other condition will be represented by a low voltage if the first one was high, or vice versa, and by definition it means that the condition is FALSE, or RESET, or LO, or OFF. Switches are the devices that select levels.

The first digital computers, made back in the 1940's, used the contacts of relays connected together for switching circuits. In the next decade, switching action in computer circuits was done by vacuum tubes, later by semiconductors and resistive elements, and more recently by transistors.

To illustrate switching, a circuit using semiconductors and resistive elements is shown in Figure 3-1. A low voltage signal at A or B causes the appropriate diode to conduct, and the output at C is as if turned off at the LO voltage level. If both A and B have high voltage level signals applied to them, both diodes are back-biased and neither conduct, so the output is as if turned on at the HI voltage level.

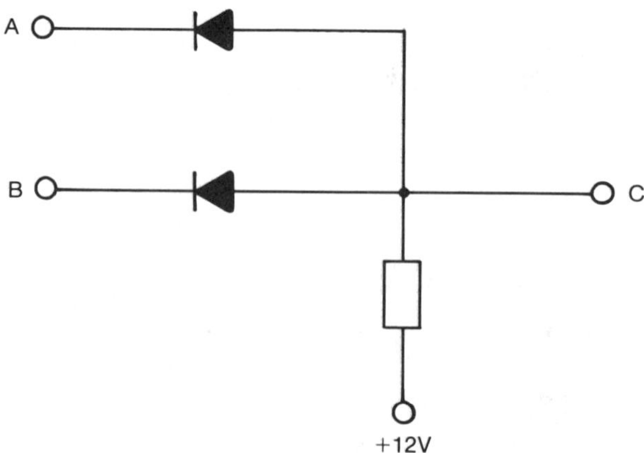

Figure 3-1. A Resistor-Diode Switch

3-3 Transistor Switches

After transistors were developed it was soon discovered that they had characteristics that made them very useful for switching circuits. Transistors are smaller than mechanical switches or relays and use less mounting space. They perform the switching operation more quickly than diodes, and their action is more precise and reproducible.

The bipolar junction transistor (BJT) and the field effect transistor (FET) are classes of transistors used extensively in computer circuitry as active devices for both switching and amplification. (A transistor, which changes the state of its output, is an active device. A resistor, through which current passes with no change in state or polarity, is a passive device).

The bipolar transistor functions as a switch because it is in one of two states. Figure 3-2 shows a bipolar junction transistor performing a switching operation. An input voltage V1 of sufficient magnitude applied to the base makes the transistor conduct; when there is no signal, there is no conduction. When it conducts, current through the LED causes it to light. A relatively small drive voltage, such as the HI/LO signal change of another circuit element, will make the light behave exactly as though application of the input voltage had caused a switch to open and close. Table 3-1 lists the relationships.

Table 3-1
Input/Output — Resistor-Diode Switch, Figure 3-1

A	B	C
LO	LO	LO
LO	HI	LO
HI	LO	LO
HI	HI	HI

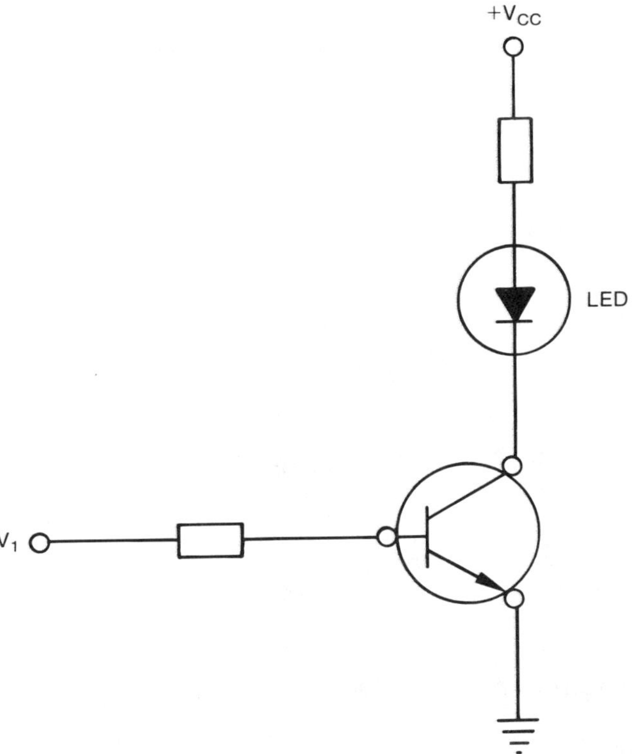

Figure 3-2. A Bipolar Junction Transistor Switch Used as an LED Indicator Driver

In the FET-type circuit, current through a single type of transistor (p or n)* is controlled by the influence of an electric field. Figure 3-3 illustrates schematically an n-type metal oxide conductor field effect transistor (MOSFET). There is no current flow from Source to Drain when the Gate voltage is below a threshold value. The resistance between Source and Drain is very high, very nearly that of an open switch. When the gate voltage is above the threshold, however, the transistor conducts, and the resistance between Source and Drain becomes very nearly like a short circuit. This makes an excellent switching device. The FET switch has advantages over the bipolar transistor: high impedance between the Source and Drain connections minimizes loading between elements in FET logic circuits; it offers high analog transmission accuracy because it does not introduce a junction potential in series with the signal; and it has simple construction, small size, and low power consumption. This last attribute is important because it results in operating temperatures that will be lower than for other types of transistor. High temperature degrades the functionality of circuitry, and ultimately limits the number of devices that can be built into a chip. However, the FET is generally slower in its switching action than the bipolar-type switch.

3-4 Logical Circuits — Level 2

Semiconductors can be combined to form logic elements called gates, used to perform switching functions. A lot of logic is done at the gate level. A gate either develops an output or does not, depending on the status of a combination of inputs. The most commonly used gates are shown as symbols in Figure 3-4. The switching circuit illustrated by Figure 3-1 gates the signals at A or B. It uses resistor-diode logic, but its action can also be accomplished by a resistor-transistor network. The resistor-diode combination is slower (if you care to call 50 nanoseconds slow) than the resistor-transistor implementation.

Gates made with transistors differ in speed, noise immunity, versatility, and durability, depending on the types of transistors used and the methods used to turn them on and off. Two commonly used variations are Transistor-Transistor Logic (TTL), and CMOS Logic. A third variation, Emitter Coupled Logic (ECL), is sometimes used.

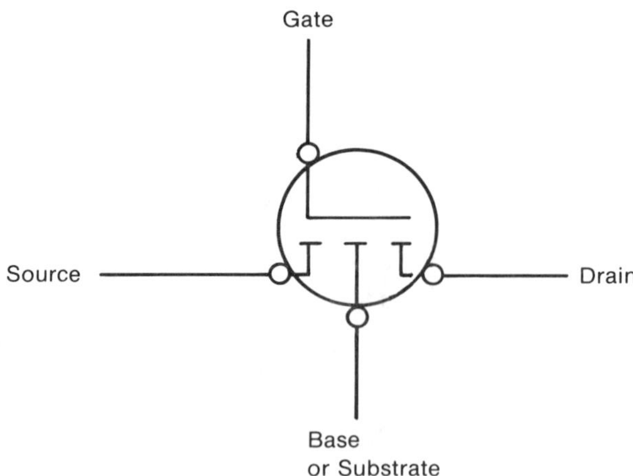

Figure 3-3. N-Type Metal Oxide Conductor Field Effect Transistor (MOSFET)

---

*For a definition of p-type and n-type junctions, see section 3-12.

TTL Logic

TTL circuits provide superior speed, power dissipation, noise immunity, and reduced propogation delay time. An example is shown in Figure 3-5.

Q1 is a transistor that acts as a gate. In this case, it acts as an OR gate because if either the A *or* the B input is at its binary 0 level, it will conduct, switching on any load that is connected to ground through the output connection.

Its operation, listed in Table 3-2, is opposite to that of the diode-resistor switch of Figure 3-1. That is, if either or both of the inputs A or B is HI, the output will be LO, equivalent to a binary 0. When both inputs are LO, the output is HI, corresponding to binary 1.

Some buses carry information on a time-shared basis, first in one direction, then in the other. There is a special form of TTL circuit called a three-state buffer, used to send data over buses that carry information in two directions. A standard TTL switch connected to such a bus may be putting out a signal because of the status of its inputs, and, even though that is not the switch with the desired information, its 1 or 0 signal might become an interference with the output of other switches connected to the same bus. TTL switches for this service are designed to have a third state available, which is essentially an open state. It corresponds to a very high impedance and is effectively equivalent to disconnecting the switch from the bus, except when its signal is the one of interest.

CMOS Logic

Gates can also be made using field effect transistors. FET gates offer the advantages of high input impedance, low power consumption, greater circuit simplicity, and reduced size. A number of FET gate types have been manufactured, including designs based on p-channel

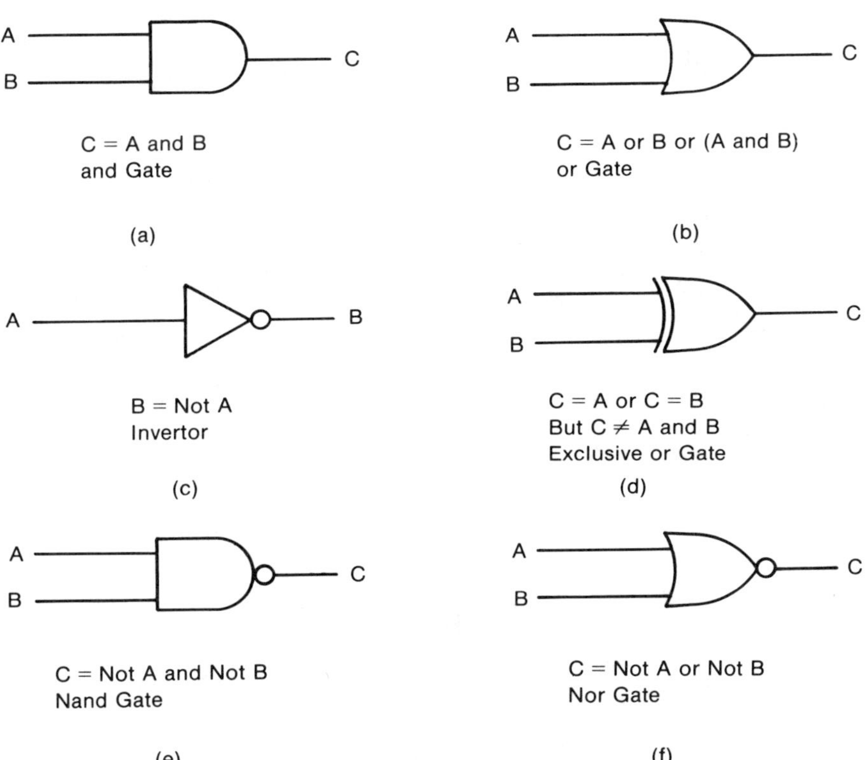

Figure 3-4. Symbols for the Most Commonly Used Gates

FET's (P-MOS) and n-channel FET's (N-MOS). They are highly successful for many LSI circuits such as microprocessors, large memories, and calculators, but none of them have displaced the TTL circuit for basic gate and flip-flop applications. A logic family made of complementary p-channel and n-channel pairs of MOSFET transistors (CMOS) is, however, widely used for basic logic operations. Among its advantages is expanded voltage source capability. CMOS logic can operate over a range of power supply voltages from 3 to 18 volts. (TTL operates over the much narrower range of 4.75 to 5.25 volts.)

A logic gate that performs functions similar to the TTL circuit of Figure 3-5 but is made from CMOS FET's is shown in Figure 3-6. It uses two p-channel FET transistors connected in series, with two n-channel FET's connected in parallel. The operation of the switch, listed in Table 3-3, is the same as that of the diode-resistor switch shown in Figure 3-1. When either or both of the inputs are HI, the output is equivalent to binary 0. Only when the inputs are both low is the output a binary 1.

Table 3-2
Input/Output — TTL Switch, Figure 3-5

A	B	C
LO	LO	HI
LO	HI	HI
HI	LO	HI
HI	HI	LO

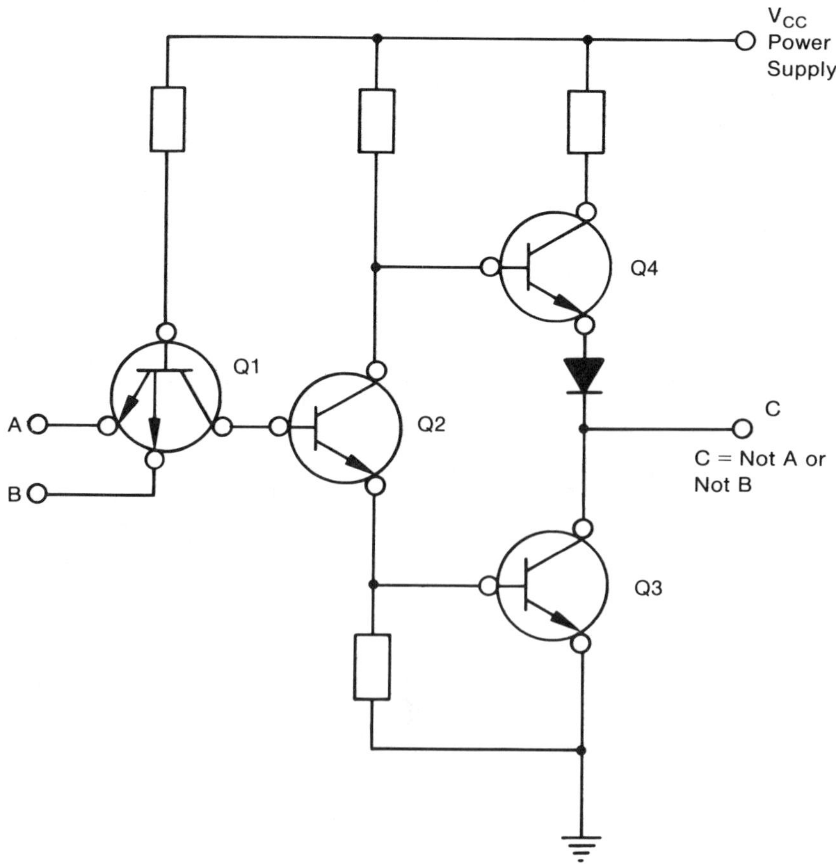

Figure 3-5. A Transistor-Transistor Logic Gate

The low heat generation characteristic of CMOS (complementary metal gate oxide insulation film semiconductor) has caused it to be used extensively in integrated circuits. Passage of current through circuit elements always generates some heat, and heat degrades a transistor's performance more quickly and significantly than any other environmental condition. Miniaturized components do not create nearly the amount of heat that their large scale counterparts do. Nevertheless, even though it may be generated by miniaturized transistors, temperatures in chips packed into a small area can reach limiting levels. Because of this, the

Table 3-3
Input/Output — MOSFET Switch, Figure 3-6

A	B	C
HI	HI	LO
HI	LO	LO
LO	HI	LO
LO	LO	HI

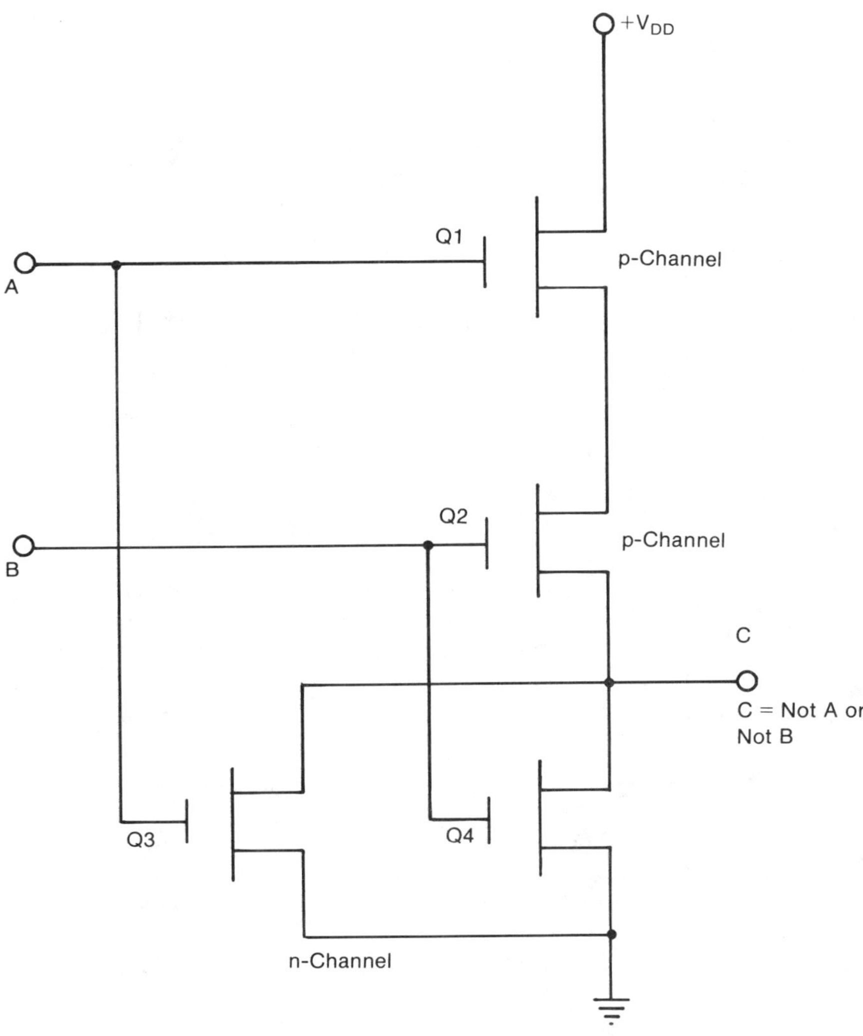

Figure 3-6. A Complementary CMOS Logic Gate

ultimate (or at least the next step) in semiconductor design may be the supercooled assemblies that promise many times the power and speed of today's semiconductors. Digital control devices will probably never require the power that such a design will provide.

Today, CMOS finds favor because its design reduces the heat generation to a minimum. This is because current passes through the circuit only when there is a transition in the state of the component. In the static state, which is most of the time, the only currents drawn are junction and subthreshold leakage currents, and these are minute. In CMOS semiconductors the output logic level transition occurs when input voltage is halfway between the drain supply voltage V_{DD} and ground. This gives excellent noise immunity and low power dissipation.

ECL Logic

The third type of logic, emitter coupled logic, is used for circuits that require very high speed switching. It uses bipolar transistors in a different mode of operation from the TTL circuitry, but at considerably higher cost and power consumption.

3-5 Registers and Latches

Switches pass information one bit at a time. At the level of complexity that deals with the transfer of information one word at a time (a group of bits that is moved as a unit is called a word; a word may be made up of eight, or sixteen, or even more bits) the subcomponents involved include latches and the assemblies that are made from latches, such as flip-flops, registers, adders, counters, and decoders.

Switching functions, however implemented, are combined in integrated circuits to form subcomponents with two basic functions. One function is information storage, the other is decision making.

3-6 Information Storage

Latches are used to store information until it is needed. Figure 3-7 shows a latch made from NAND gates. This combination is sometimes called a set-reset flip-flop. When there is a change in the state of the inputs, the outputs change to alternate states and hold that condition until the next time the input state changes. The output values for the set and reset inputs being true are shown in the table accompanying the figure.

Combined in groups to form registers, a flip-flop can be used to hold a series of bit values. A succession of flip-flops can be used as a shift register, where bit values are shifted from one latching unit to the next; or as storage registers, where bit values are held until called for.

There are variations of flip-flops, but in general terms they are a form of latch, a combination of switches that assumes a condition when one input is energized and another condition when a second input is energized. They are analogous to a two-coil electromechanical relay: energizing one coil opens the relay contacts; energizing the other coil closes the contacts. Flip-flops triggered (or "strobed") by clock pulses, are basic building blocks for registers and counters.

Figure 3-8 illustrates a type of flip-flop that is used for the units of registers. This is called a D flip-flop. It has two outputs: if the $\overline{Q}$ (Not Q) output is high, it indicates that the contents of the flip-flop is a binary 1; the Q output indicates the opposite, the presence of a binary 0. The T input is the strobe line and activates the unit. Since it connects to both AND gates, nothing will happen unless it is high. The two OR gates form a latch where the bit is stored.

3-7 Memory-Storage Registers

Memory can be thought of as a large array of mailboxes, each containing a number of latches. Eight latches can represent an eight-bit word. Those latches holding binary 1 level bits correspond to switches that are closed. Those holding values representing binary 0's correspond to open switches. Even the number of the mailbox is represented by a group of switches. Suppose that mailbox number 10011100 holds the word 00111110, designated by a bit arrangement or combination of on-and-off switches. If a command in a program directs that the contents of the mailbox represented by 10011100 be moved into the mailbox with the number 11111110, then 00111110 will be moved over an eight-foil bus to the register that has the address 11111110.

Figure 3-9 illustrates one cell of memory. When the cell is selected for reading, the stored value appears as a HI or LO at the Data Out pin. In memory types where the content is alterable, the Write Select input allows the cell to take on a 1 or a 0 value, whichever is on the input bus.

Memory is organized in arrays of bytes and will be either Read Only Memory (ROM) or Random Access Memory (RAM). RAM includes circuitry that permits bit patterns to be changed at the direction of a programmed instruction. When the desired bit pattern is presented to the data input and the *write* clock input is pulsed, the bit pattern will be loaded into the register and its "switches" set to hold that pattern. It can also be read out of, so that the information in the "switches" can be accessed and moved to another register.

ROM, on the other hand, can only be read from; the user cannot change it. ROM is stable. Like a permanent magnet, it retains its information even when there is no instrument power. RAM, on the other hand, is volatile. Its information will usually be lost if power fails.

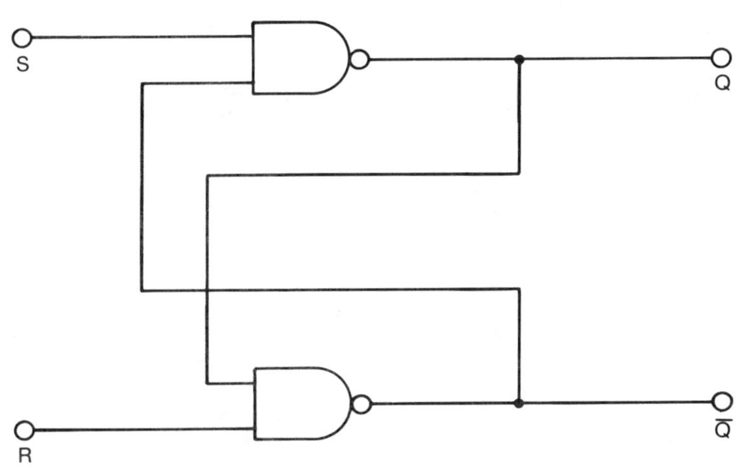

Inputs		Outputs		State
S	R	Q	Q̄	
0	0	1	1	—
0	1	1	0	Set
1	0	0	1	Reset
1	1	X	X̄	Either Set or Reset

X is either 1 or 0, depending on previous input conditions

Figure 3-7. NAND Gate Latch

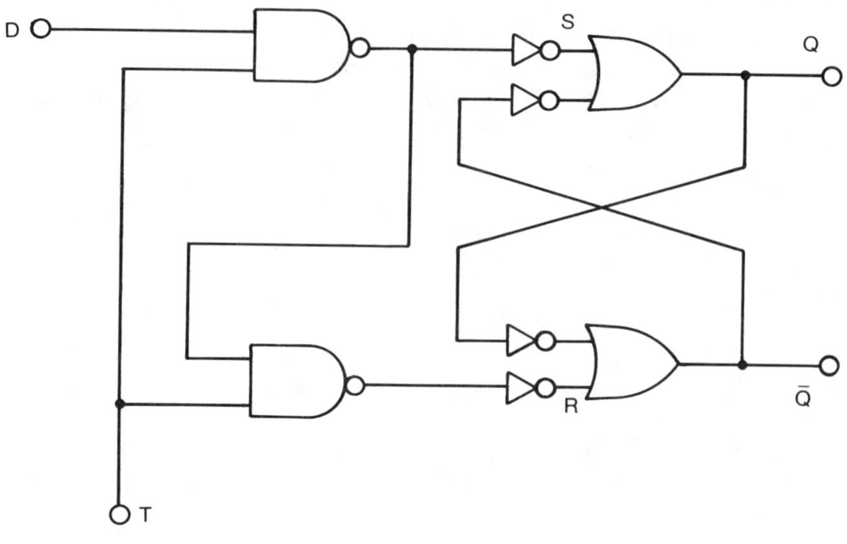

Inputs		Outputs	
D	T	Q	$\overline{Q}$
0	0	X	$\overline{X}$
0	1	0	1
1	0	X	$\overline{X}$
1	1	1	0

Figure 3-8. A Type D Flip-Flop

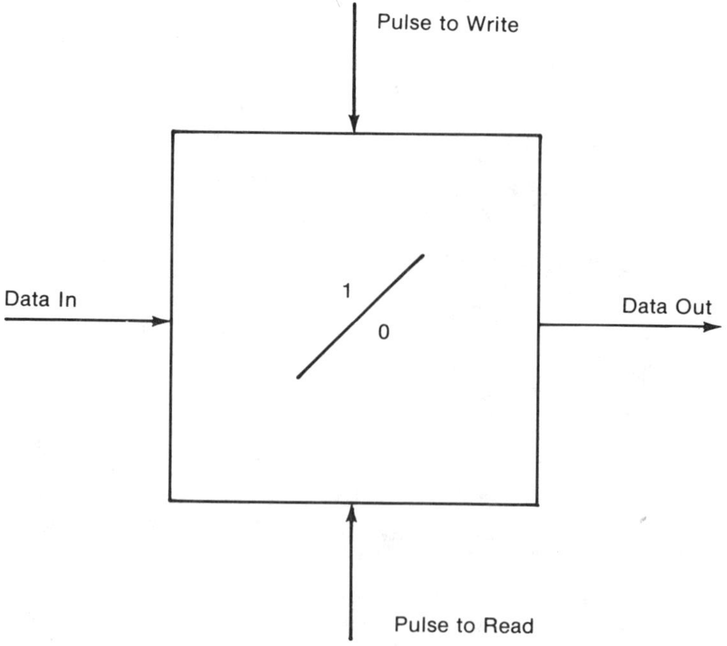

Figure 3-9. A Memory Cell

3-8 Shift Registers

The shift register is a sequential logic device made of binary storage elements, usually flip-flops, cascaded in a way that allows the contents of one to move into the next when a pulse from a clock activates it. All the segments of the register are strobed at once by a clock pulse, and so the bits all move together, as shown in Figure 3-10.

Bits input serially are shifted from one flip-flop to the next. The type of flip-flop used is called a JK flip-flop, which differs from the D type flip flop in that it is triggered by an input change from 1 to 0, but not from 0 to 1. After four clock pulses, four bits have been shifted into the register, displacing four bits from the other end into the adjacent register with the next sequential address. In this way, a message of many words can be stored into a block of memory. Circuits can also be put together that will pass memory into and out of registers in parallel.

Many programmed operations take place in shift registers. A few examples follow, illustrating information transfer.

(1) To take a serial string of bits from a bus and put it onto 8 bus lines in parallel, shift the bits one at a time into a shift register, then take them out all at once onto the parallel bus.

(2) Multiplication by 2 is accomplished by shifting a 0 into the register holding a number. 0101, decimal 5, becomes 01010, decimal 10, when a 0 is shifted into the least significant bit position.

(3) To examine the state of the third least significant bit in the word 00110101, it is shifted right three times, to become 00000110. If the bits are shifted out into another register, the 1 that was in the the third position can now be isolated (00000110 1) and examined to see if it is a 1 or a 0.

3-9 Counters

Another sequential logic device made from the JK flip-flop is the counter, used to count the number of binary logic level transitions applied to it. Figure 3-11 illustrates such a counter.

The operation can be followed by examining the wave form patterns in the sketch. The description starts with all units containing logical 0's, so that the number stored is 0000.

The first input transition from 0 to 1 puts a 1 into the first flip-flop; transition back to 1 does nothing. The number stored is now 0001, equivalent to decimal 1.

The next pulse toggles the first unit, and it resets it to contain a 0. As it resets from 1 to 0, it toggles the second unit, so that it contains a 1. The number stored is now 0010, equivalent to decimal 2.

The third input pulse will reset the first flip-flop, changing its contents from 0 to 1. This does nothing for the second unit, so the stored number is now 0011, or decimal 3.

The fourth input pulse resets unit 1 to 0, and this transition resets number 2 to 0. This transition in turn toggles number 3 to make it latch in a 1, and now the stored number reads 0010, or decimal 4.

```
1 0 1 0  | 0 | 1 | 1 | 0 | 0 | 0 | 0 | 1 |     AT THE START PULSE

  1 0 1  | 0 | 0 | 1 | 1 | 0 | 0 | 0 | 0 |     AFTER FIRST PULSE

    1 0  | 1 | 0 | 0 | 1 | 1 | 0 | 0 | 0 |     AFTER SECOND PULSE

      1  | 0 | 1 | 0 | 0 | 1 | 1 | 0 | 0 |     AFTER THIRD PULSE

         | 1 | 0 | 1 | 0 | 0 | 1 | 1 | 0 |     AFTER FOURTH PULSE
```

Figure 3-10. Bits Moving through a Shift Register

3-10 Decision-Making

The second important function of registers is to provide a means for making decisions. Depending on the state of a bit, or the composition of a word, the sequence of programmed instructions may be required to jump to another part of the program, loop back to a previous command, or branch to a different sequence.

A function that uses registers and occurs at this level is addition of bits. All logical arithmetic is based on addition. By moving bit arrangements into registers and adding other bit arrangements from other registers, comparisons can be made by addition. The net result is that bit arrangements can be manipulated, commands can be performed, and programs implemented.

The addition of two bits can be accomplished by using an exclusive OR gate. Recall that the exclusive OR develops an output if, and only if, one of two inputs is true. If both, or neither, are true, the output will be false. Bit addition is as follows:

$$0 + 0 = 0 \qquad 1 + 0 = 1 \qquad 1 + 1 = 0 \text{ and carry } 1$$

The exclusive OR gate operates exactly the same way, except that there is no carry bit generated when both inputs are true. To accomplish this, an AND gate is added to form a combination called a half adder. See Figure 3-12(a). This still does not accommodate a carry bit from a less significant bit position, but combining two half adders will solve that problem. See Figure 3-12(b). If a full adder is supplied for each bit position, addition of two registers of bits can be performed.

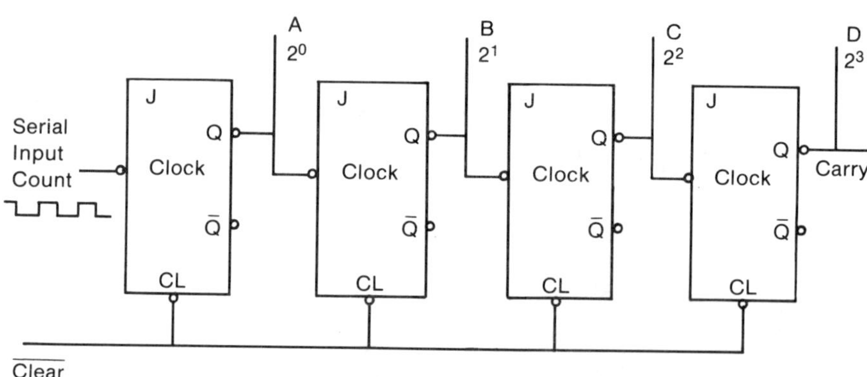

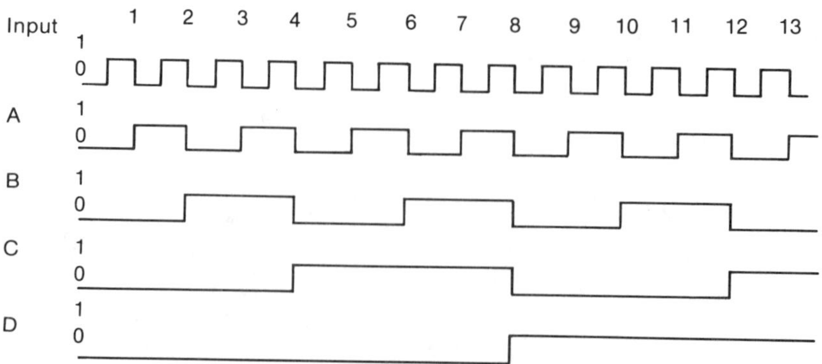

Figure 3-11. Binary Counter

The exclusive OR gate is not the only function used to make decisions, but it is convenient and is used a great deal. Some additional examples of its usefulness follow.

(1) The exclusive OR gate can be used to fill a register with 0's, a step occurring during initialization of a program. Adding a number to itself without any carry can be done by this gate to get a guaranteed 0 in every position.

(2) Another chip operation uses this gate to check for the state of any specific bit position in a register. In the section above it was explained that the state of a bit could be checked by moving bits one at a time from a shift register and examining each one to see if there is a match with a 1 or with a 0. The examination is done by a comparator. A common way to compare digital values is to XOR (exclusive OR) them. If both are the same, the result will be all 0's. If they are not the same, the result will not be all zeros, and the condition can be detected, indicating a mismatch. Comparators are used in conversion systems to balance input signals against internally produced references.

This explanation may have seemed to be excessive for a text that is not intended as a design text. It has been done deliberately, at the risk of losing some readers' attention, to make the point that digital logic is nothing more than combinations of very simple functions, accomplished by moving bits of information from register to register. There are many, many common gate combinations, and examples are given for only a few. All of these combinations, or, more correctly, all of the active and passive elements making them up, can be obtained as parts of standard chips.

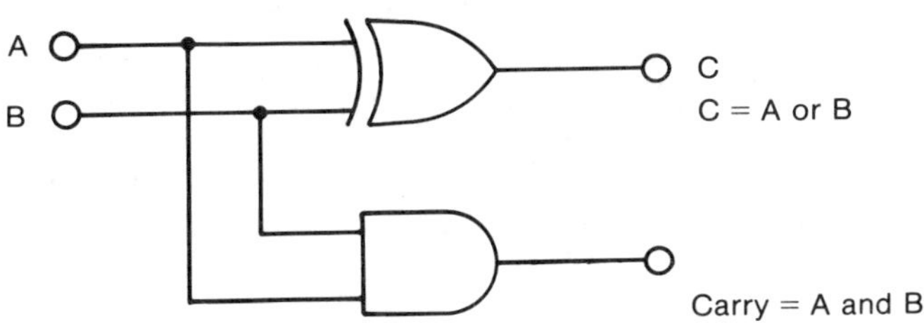

(a) Half Adder

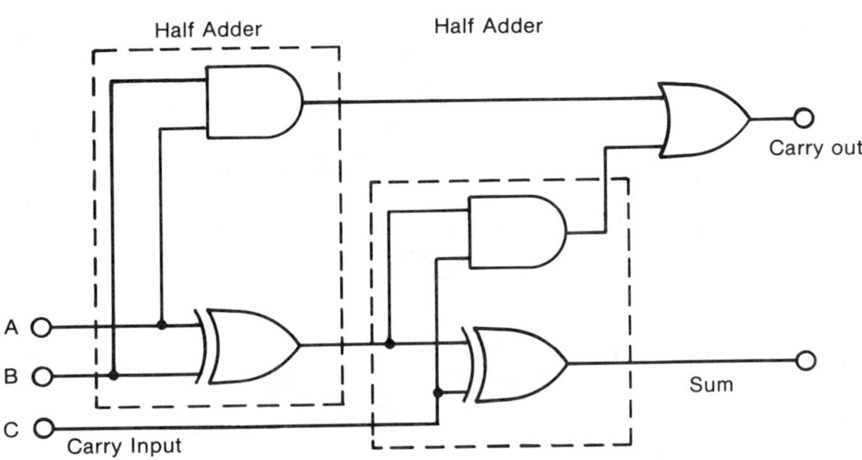

(b) Full Adder with Carry

Figure 3-12. Binary Adder

3-11 Decoders and Displays

Subcomponents can be formed by combining registers and counters. Decoders and displays are examples of this design class.

Decoders are used to tell an integrated circuit what to do. In order that a device addressed knows what a command means, the bit pattern of the command must be decoded in such a way that activating signals arrive at the pins of the device to initiate the desired action.

Commands are transmitted in the form of bit patterns. If the bit pattern 0010000 is sent to the pins of a chip designed so that the pin receiving the 1 input tells the chip circuit to read information, its latches will be set to do just that. On the other hand, if a 0 is the input that determines a write operation, this will occur if the bit in the third position is set to 0 level.

Figure 3-13 illustrates the action of a 2-bit binary decoder. The circuit decodes the four possible states of the 2-bit word AB. It is known as a 2-line to 4-line decoder.

Several of these subcomponent circuits can be combined to perform a more sophisticated task. Figure 3-14 puts together a counter, latch, decoder, and display for single decade counting. A pulse to the latch enable sets four bits of the latch to the count value. The BCD digit at the latch output is decoded to drive the appropriate segments of the display. The counter must advance by one count every time a valid pulse appears at the input.

3-12 Integrated Circuits

Today, switching devices are built into silicon substrates (a substrate is a base material into which other materials are diffused, or inlaid, or etched). The combination of circuits is commercially available as a chip. Since many of the third level of digital logic components occur as chips, or as significant portions of the components contained in a chip, a discussion of the physical construction of integrated circuits will introduce this level of circuit design.

The integrated circuit phenomenon started with small scale integration (SSI), progressed through medium scale integration (MSI) and large scale integration (LSI), and is now at the stage of very large scale integration (VLSI), as more and more circuits and components are crowded onto a single chip.

All digital devices today use integrated circuits, This development has reduced the size and price of computers to the point where they have become available to everyone. Because of integrated circuitry, costs of digital equipment have fallen significantly and the boundaries of design have been extended. The lower cost comes about partly because the very basic circuits can be used repeatedly in a multitude of combinations, a form of mass production.

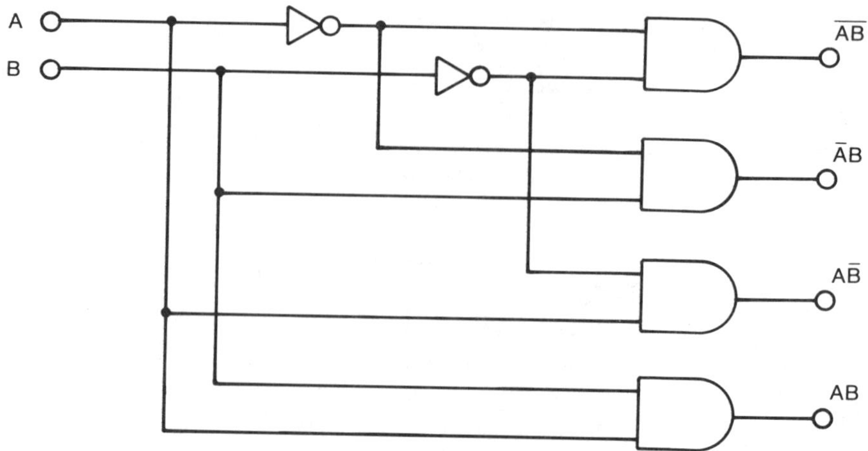

Figure 3-13. Two-Bit Binary Decoder

Our economy develops to a large extent from the fact that mass production is cost effective, and this is borne out in the case of electronics by the fact that the cost of integrated circuits has fallen so low that in many cases they contribute only 10% to the total cost of a product operating from electronic circuits.

Thousands of repetitions of the same basic devices can be reproduced on a single wafer of silicon. If large numbers of the same circuit can be used over and over again, integrated circuits can be made very economically. Fortunately, the basic circuits — very elementary ones indeed — can be used in many different applications throughout industry.

How is this done using a thin slice of silicon?

Semiconductors are produced by deliberately introducing impurities into silicon crystals. Pure silicon is not a good conductor; it is chemically quite inert. Doped with minute amounts of a metal such as antimony, however, free electrons become available, and a semiconductor is formed that has the ability to pass a negative current. This is called an n-type semiconductor. If an element such as indium is used instead of antimony, a p-type junction is formed, and the doped silicon can pass a positive current.

Large scale integrated (LSI) chips are silicon wafers (called substrates) in which doped areas are created and interconnected. The interconnection and fabrication is a complicated many-step process of photoengraving, masking, etching, diffusion, and overlaying. The many steps in the complicated process of manufacturing a chip would seem to make production very expensive, but this is not the case. Mass production and mechanization reduce the cost of the complex process. These fabrication processes result in a chip that has, literally built into it, the electrical equivalent of transistors, resistors, capacitors, and, of course, switches. Figure 3-15 shows the relationship of doped areas in a representation of a chip.

Tiny wires connect the "components" on the layers of substrate to pin extensions that bring signals into and out of the circuitry. The assembly is imbedded into a protective covering with the pins protruding, and a "chip" has been created. The pins are spaced so that

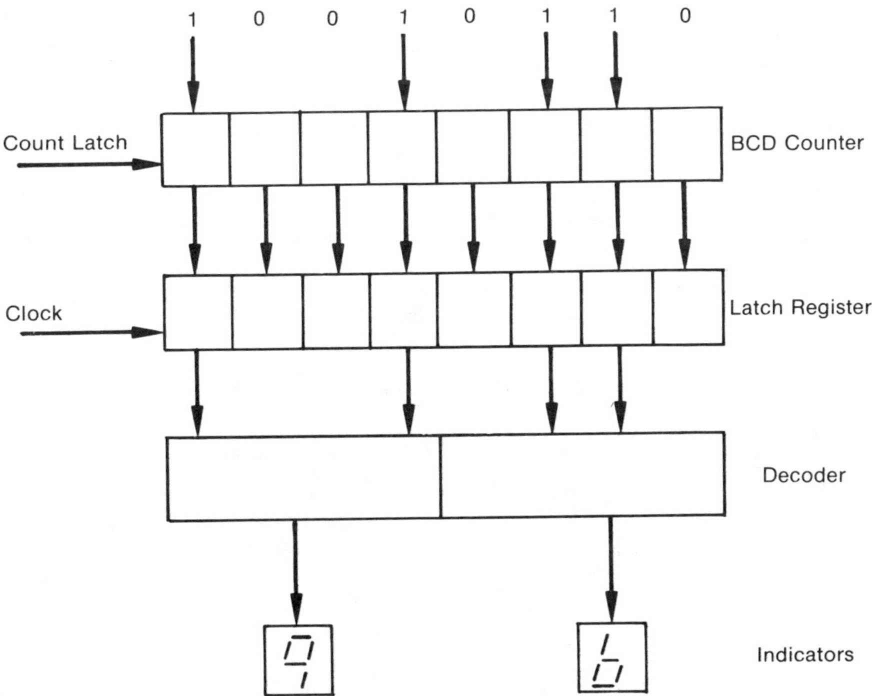

Figure 3-14. Counter, Latch, and Decoder Combined to Actuate an Indicator

they can be pushed into a socket mounted on a printed circuit board to become part of a digital control device or assembly.

A large scale integration (LSI) device can be thought of as a combination of many gates, connected together with passive elements (resistors, capacitors, and inductors) to function as switches performing logical operations. It is possible to construct many passive elements in a silicon substrate, controlling their values by changing the shape and size of etched areas. LSI puts many standard designs together on a single chip to produce other more sophisticated but still standard designs.

3-13 Clocks

Operation of many devices using combinations of counters, shift registers, latches, and decoders depends on precise timing. This can be provided by a clock, a device that produces a periodic signal that can be used to step sequential circuits through their operating procedure.

Clock signals can be generated by an oscillator, using any combination of components that turn each other on and off. Another type of clock signal is produced by a one-shot, or monostable multivibrator. This device develops a fixed duration pulse each time it is strobed. If the trigger pulse comes from an external crystal source, very accurate clocking results.

3-14 Logical Circuits — Level 3

Up to this point, we have been talking about the circuit building blocks. The assemblies discussed in the remainder of this unit are put together from the primary subcomponents (switches, flip-flops, decoders, etc.). At the next level of complexity are circuit components that perform a function necessary to the processing of information. The basic ingredients are combinations of the components discussed previously, with a precise time base and involving programming. Programming schedules the transfers of bit patterns in sequences that satisfy operating requirements.

These components are much more sophisticated than the circuits previously described, but they will still be considered to be basic. They are necessary to, and are used by, every type of digital control device. They include microprocessors, A/D and D/A converters, multiplexers, 7-segment displays, etc. A few years ago they would have been supplied as complete printed circuit cards. Today, they will probably be manufactured as a chip or part of a chip.

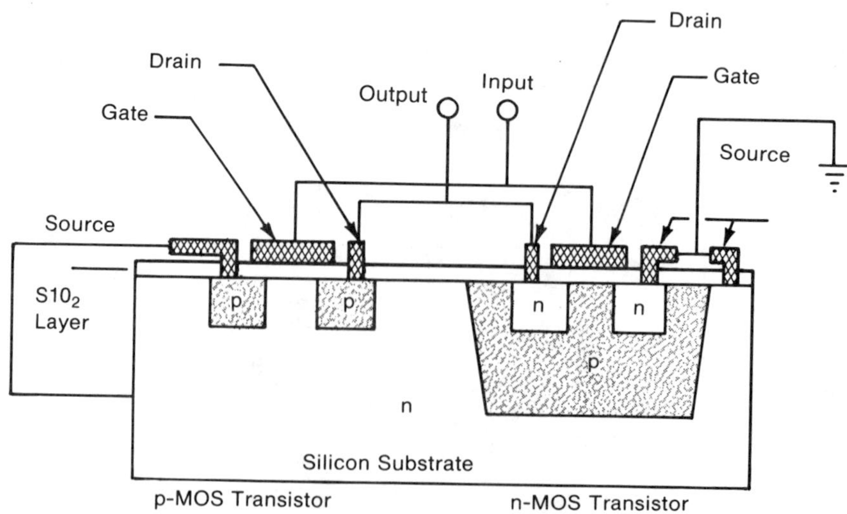

Figure 3-15. Complementary MOS Solid State Device

3-15 The Microprocessor

The microprocessor is the heart of the digital circuitry, yet it is nothing more than a combination of switches making up general and special-purpose registers with a system for controlling data flow and register operations, paced by a clock that also aids in timing communication with external devices. The reason it is so special, of course, is the use that is made of the switches and registers through programming.

A story of espionage, written by an author with some knowledge of information flow through a computer and a lot of imagination, used the numbers on the box cars of a complicated model railway system to store secret records by moving the cars around and combining them in staging yards in various combinations. The real life equivalent, a railroad terminal dedicated to handling freight and commuter traffic, is an analogy of the way a microprocessor does its job. Freight, for example, is stored in warehouses and on freight cars located on track sections dedicated to storage areas, or is entering or leaving the yards as part of made-up trains. Passengers wait in waiting rooms or on platforms until their trains are announced, then board the cars and are taken to their destinations. Other passengers arrive on incoming trains and either leave the station, exiting to the street or to other buildings, or sit in waiting rooms until their connections are announced. Cars from incoming trains are shunted to tracks where they are combined with other cars to make up new train combinations, in accordance to the commands from a dispatcher's office. Trains with the proper complement of cars, freight, and passengers are assigned to an outgoing track and sent on their way. All of this happens in accordance with instructions, a program of operations that schedules arrivals and departures, movement of freight, preparation of manifests, and the storage of people and equipment.

Upstairs in the Operations Division of the business offices, the programs are made up and implemented. This area is equivalent to the Central Processing Unit (CPU) of the microprocessor. The activities of the CPU are also governed by a program, an executive program. It operates in accordance with other programs, subroutines that define payment of personnel, that define the kinds of cars to be used on each type of train, that maintain the station and the equipment, and hundreds of other operating procedures that determine the organization of the entire railway system. With this analogy in mind, the operation of the microprocessor, illustrated in Figure 3-16, should be evident.

Information is moved between the registers and memory, in both directions. The waiting rooms and warehouses, along with the roundhouse and assembly yards, are registers making up memory. The track, of course, is a bus. The information will be in combinations of bits — 8, 12, or 16 are the usual, but not the only, sizes of the words, depending on the brand of microprocessor used in the instrument under consideration. As might be expected, 32-bit processors are beginning to be used. The information will represent real numbers, or memory addresses, or the contents of memory locations. Program commands that determine what to do next, data, and addresses, come from RAM, arranged in schedules prepared in the Dispatch Office from instructions sent by the Operating Division. Instructions that define the microprocessor operation, such as how to read the keyboard input and how to respond to interruptions, are stored in ROM, the Business Office operations manual. The information moves over three separate buses: commands over a control bus, teletyped to the dispatcher's office; data to and from memory over an information bus; individual car movements and instructions that define where to find information and where to store it, over an address bus from the Dispatch Office.

The most important register (or more likely, group of registers) is called an Arithmetic Logic Unit (ALU). Informaton to be examined and/or processed is moved into the ALU from storage or from a working area reserved in memory, a temporary stack of memory locations. Movement is in response to a program, a sequence of commands. One special register may be assigned to keep track of the program sequence. It counts the steps and has a pointer to each program step address, so it knows where to call the next instruction from.

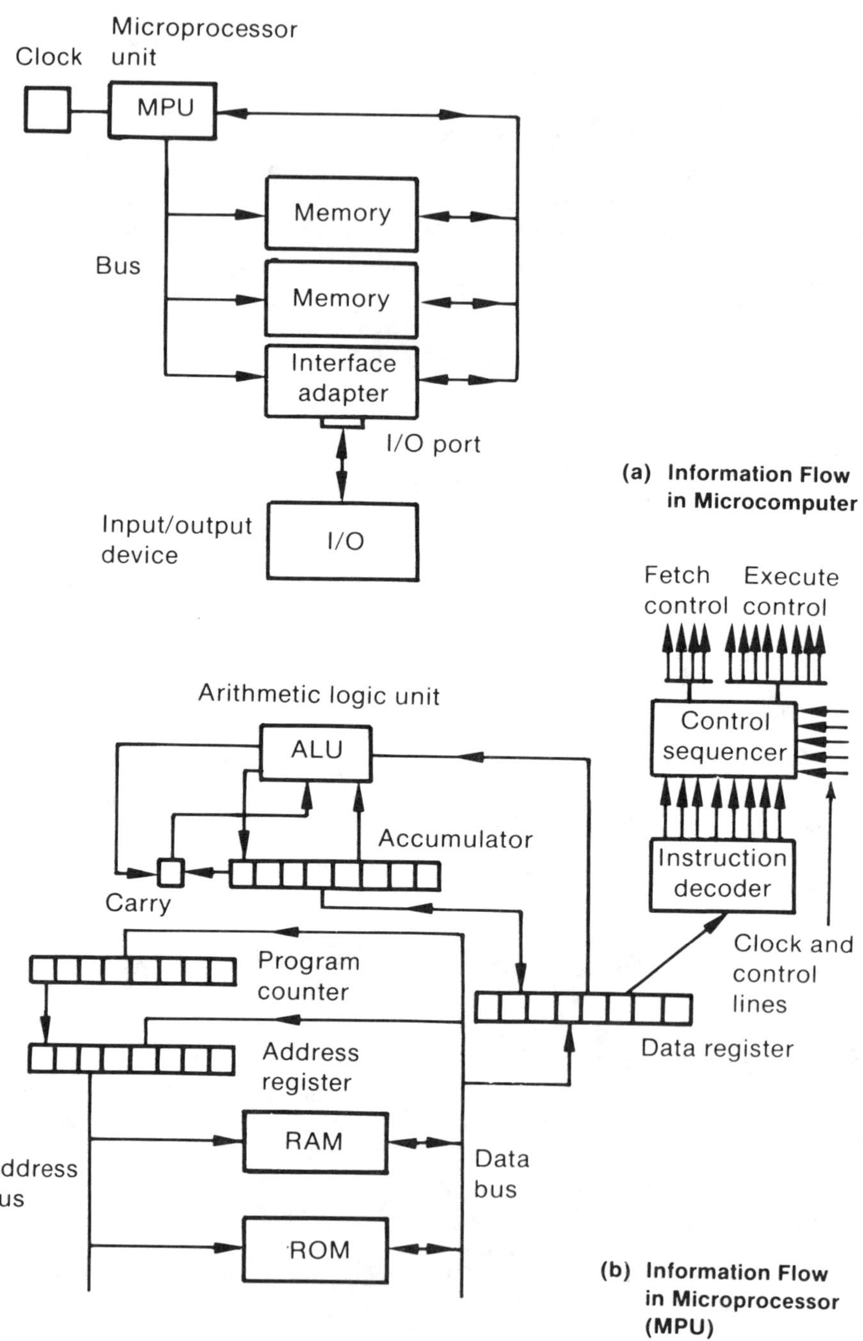

Figure 3-16. Microprocessor and Arithmetic Logic Unit

(From *Understanding Distributed Process Control*,
J. A. Moore and S. M. Herb, ©1983, Instrument Society of America.)

The majority of commands will be for different forms of data transfer, but there are other commands for arithmentic operations (ADD, SQRT), logical operations (AND, OR), Branch and Call, Respond to Interrupts. Call instructions create a branch to entire subprograms, sequences designed to perform special operations like multiplication or indexing to specific areas of memory. All of this information transfer, moving the necessary information in the correct sequence, is the essence of computer operation. This is true for a single microprocessor or for a complete mainframe computer, and for all of the digital control devices in between.

Microprocessors may be packaged on a single chip, often (for 8-bit or 16-bit computers) with 40 pins. These provide connections for a clock input, power, Read/Write instructions, reset, and buses for addressing memory, data, and control signals.

3-16 Signal Conditioning

A very important operation, necessary to interface the analog signals from field-mounted transmitters to the binary format of a computer and at the same time keep out false information, is signal conditioning. Input signals from process-measuring devices will not be suitable for digital processing until they have been converted to a form and voltage level that the processor can accept. All signals, analog as well as discrete, have to be conditioned, but the requirements for the two categories are not the same.

Sometimes some of the requirements may be fulfilled by components external to the digital control device. For that reason, some of the considerations of signal conditioning will be discussed further in Unit 4 which deals with systems. All necessary functions, however, may be fulfilled in the solid state circuitry itself, or by peripheral components connected to it within the housing of the digital control device. Accordingly, the major review of this subject will be in this section.

3-17 Conditioning Digital Inputs

Digital inputs come from mechanical contact closures, electronic switches such as proximity switches, and electrical pulse trains. Voltages across mechanical contacts may be ac or dc, and may range from 5 volts to 220 volts. Electronic switches switch a narrower range of voltage but need a return path to keep the sensor in an OFF state. Pulse train signals may come from flowmeters, tachometers, or from minicomputers and single-loop controllers. In many cases the pulse rate may be faster than the sampling rate of the device measuring it. If this is so, the pulses will have to be accumulated and divided by the number of pulses in a sampling period, computing an average rate for the period.

TTL circuits are often used to process digital inputs. 0 to 5 V dc is a satisfactory input level to TTL circuits, and higher voltage signals will be reduced to this level. Many digital input systems include optical isolation between the process side of the signal path and the electronic system bus. This blocks interference from external sources, preserving the integrity of the bus, which interfaces with every portion of the computer circuitry. Filtering is needed to remove the effects of contact bounce and transient voltages on the input signal lines. Some hysteresis should be provided to remove false signals from slow changing inputs.

3-18 Conditioning Analog Signals

Analog signals from transmitters measuring process values will usually be currents in a 4 to 20 milliampere range. They may, however, be voltages; sometimes these will be in a millivolt range, sometimes in a 1 to 5 volt dc or 0 to 10 volt dc range. Another source can be trains of pulses, mentioned in Section 3-17, but this time with variable spacing to represent analog values.

Most input circuits are designed to receive 1 to 5 volts, which can be readily obtained from a 4 to 40 mA signal by passing it through a 250-ohm resistor. Millivolt range signals will have to be amplified; pulse train signals will have to be accumulated and then converted to a 1 to 5 volt range. As in the case of the digital inputs, these conversions are often made external to the digital control device, but they can be done internally.

Many transmitter inputs are not linear with respect to the range of the process variable they are measuring. Linearity is *desirable* for indication of the variable and *absolutely necessary* if the signals are going to be used in computations. If linearization is done in the solid state circuitry and not by a separate external analog device, the digitized signal (after analog-to-digital conversion) may be linearized by a software program.

Signal conditioning subcomponents will include (1) some sort of input amplifier, with programmable gain, to change the incoming signals to the level required by the specific device; (2) filtering; and (3) interchannel isolation.

Analog input signals may include components or harmonics of unwanted noise, and filters should be included in the input circuitry to average the signal values. Filters can be made with hardware, such as resistor-capacitor combinations, or filtering can be accomplished by programming — digital filtering using software. Hardware filters can be low pass, high pass, or bandpass types. Low pass filters allow a group of frequencies, up to a cutoff frequency, to pass with little attenuation. Higher frequencies above the cutoff point are attenuated. High pass filters perform oppositely, passing frequencies above a cutoff point. Bandpass filtering attenuates everything below a low cutoff point and everything above a high cutoff point, allowing those in between to pass. For industrial process signal conversion it is advisable to have a separate low pass filter for each input to a multiplexed analog-to-digital conversion section.

Digital filtering has essentially no limits. Each output of a digital filter can be the weighted sum of 128 or more previous input samples. It is possible to produce very sharp cutoff points using digital filters.

3-19 Multiplexers

A digital control device may read, convert, and store many inputs. It will scan the storage registers and use only one input value at a time. A multiplexer is a component that selects one from all of the stored values and routes the selected value to a single output. For electrical circuits, a multi-pole selector switch is the most common form of multiplexer. A simple implementation for the function of a two-pole selector switch using digital logic is shown in Figure 3-17. A flip-flop sends a signal to AND gate 1 or to AND gate 2. A true input No. 1 signal value *and* the gating input from port Q, *or* a true input No. 2 signal value *and* the gating signal from port Q causes the selected signal to appear at the output of the OR gate.

A multiplexer is used to connect input signals into an A/D converter in a predetermined sequence when many independent signals must be processed by the same computer or communications channel. Multiplexing could be done while the signals are still in analog form, but digital muliplexing cuts cost to a minimum, particularly if the selection can be made locally, with only one pair of wires carrying the selected signal to a central location. Wide ranges between channels are hard to handle with analog multiplexers but are simply a matter of programming for digital multiplexers. Physical location also influences the choice. Analog multiplexing is limited to 100 meters from the converter, because the signal is prone to loss and interference. For the digital implementation, there is no limit.

A digital multiplexer extends this circuit concept, using a strobe signal to instruct individual buffers to access (read or write) a common bus line.

The design considerations for multiplexers are very similar to those for A/D converter circuits (see Section 3-21). Resolution of measurement is important, because cost increases as resolution increases. Speed of measurement is a consideration because throughput rate is a limiting parameter.

It is absolutely necessary that only one sampled value at a time be connected to the multiplexing circuit, uncontaminated by transients, common mode voltages, or leakage from adjacent input circuits. Figure 3-18 illustrates the concept of isolation as it is sometimes achieved with the "flying capacitor". Each input signal is connected across a capacitor by reed switch contacts, and the capacitor is always charged to the input signal value. When it becomes a signal's turn to be sampled, the switches associated with all signals are disconnected from the multiplexer input. Then, and only then, the switches for the signal to be selected are activated. For an instant, the capacitor is connected to nothing, then the contacts close onto the multiplexer connections. The multiplexer can never be connected to more than one signal circuit at a time.

Even more effective, but more expensive to implement, is a design that has the front end completely floating and shielded, with the guard shield completely isolated from the chassis and grounded only at the sensor. All circuits in the signal path are sheathed in a shield that is at sensor ground potential.

3-20 Sample and Hold Circuits

Sample and hold circuits must be used when the original signal varies rapidly or when a number of signals must be scanned. Any signal from a fast changing process will change during digitizing. By using a sample and hold circuit, the same A/D converter can process conversion of many inputs.

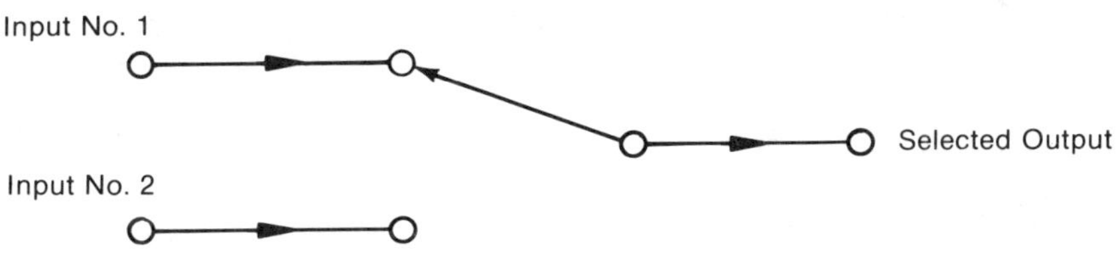

Single Pole Double Throw Switch

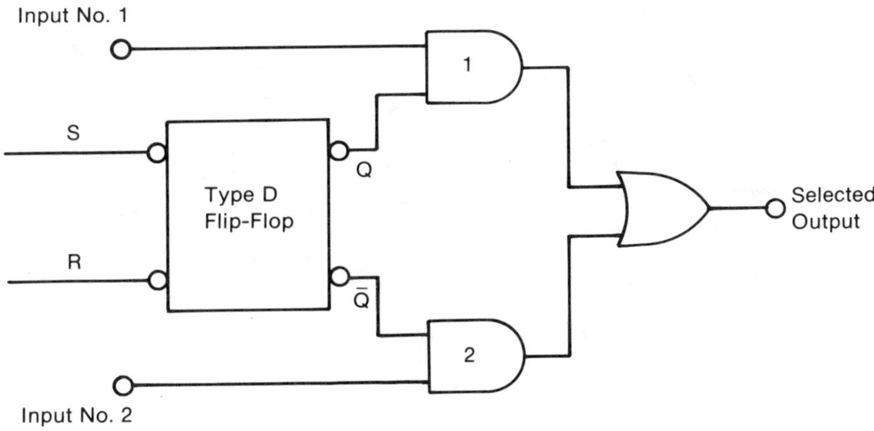

Digital Selector

Figure 3-17. Signal Multiplexing

Depending on how rapidly the process it represents is changing, the interval of reading may be once every few milliseconds (a machine tool position, perhaps), or as infrequently as once every 5 seconds (for a slow process like a steel heating furnace). The system interpolates between readings, and the interpolation is most accurate for high rates of sampling. This may create problems in control of a very fast process. For very sophisticated computer controlled systems, sampling rates have been tailored to suit changing process conditions to create a sampled data system with a variable pulse width and variable sampling intervals. In the case of processes where transportation lag exists, this may produce better results than continuous sampling. A general rule is that rate of sampling must be at least twice the highest frequency component in the signal.

Figure 3-19(a) illustrates a typical analog method of implementing the sample and hold circuit. In the sample mode the input signal charges capacitor C through the resistor R; in the hold mode the voltage on the capacitor is maintained (held) at the follower input. The follower buffers the analog signal from the low input impedance of the RC network.

A digital circuit equivalent is shown in Figure 3-19(b). A two-position switch, which can be fashioned from an OR gate and an AND gate, is connected so that it will pass a signal until switched, at which point it will select its own output and "hold" it until the switching action returns it to pass the input signal value.

3-21 Analog-to-Digital Converters

Analog signals must be represented in binary notation for use in a digital system. There are many examples of analog-to-digital conversion, and it is perhaps the most universal operation, after the microprocessor, in digital control devices.

Continuous signal wave forms can be represented digitally by periodically sampling the signal voltage and performing the conversion for each sample. The signal representation is a sequence of binary numbers. Conversion is usually to a 12-bit number, which has a high resolution, 1 mV for a 2.048-volt input. Thirteen-bit and even fourteen-bit numbers are sometimes used.

The key characteristics of an analog-to-digital converter include absolute and relative accuracy, linearity, monotonicity (does it change in one direction at all times, or is it sometimes increasing and sometimes decreasing), resolution, and conversion speed. Some sort of timer/clock will regulate the operation of the device. The functions of an A/D converter include signal conditioning, multiplexing, sampling, and conversion.

A number of standard designs are used for digital-to-analog conversion. Detailed discussion of all designs can be found in textbooks specializing in digital circuit design. As examples of types used for digital control devices, brief descriptions of several of these methods follow.

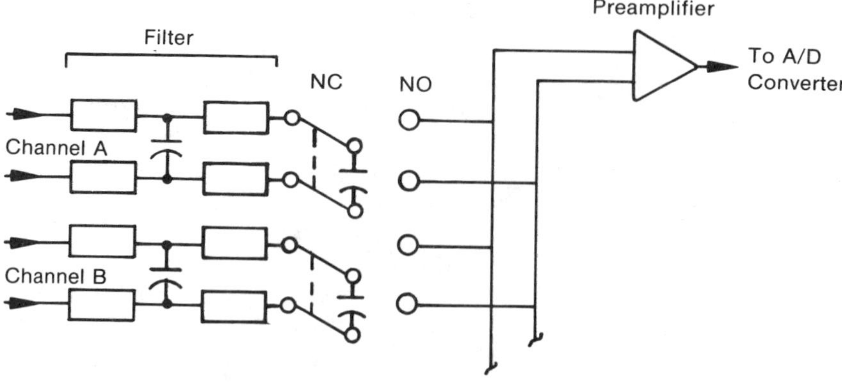

Figure 3-18. Flying Capacitor-Multiplexer

(1) *Dual slope method*. This is a three-step operation (see Figure 3-20). For the first step, an integrator is initialized to zero charge. Next, the input voltage is measured for a fixed number of clock pulses, long enough for the capacitor to charge to as high a value as it can, proportional to the input analog voltage signal. Finally, the integrator connection is switched to a reference voltage of opposite polarity. The number of clock pulses counted while the capacitor is discharged to zero establishes a digital value for the input signal.

The dual slope integration makes many errors self-cancelling in the two measuring steps. This improves its accuracy. The requirement to perform the two integrations consumes time that cannot be reduced, and so its scan rate is slower than for other methods.

(2) *Analog-to-frequency-to-digital conversion*. The analog signal is converted to a series of square wave pulses by a chopper stabilized circuit. The frequency corresponding to the number of pulses in a specific time period is proportional to the analog value. Counting the cycles during a clock period produces a binary value. Increasing the clock period increases the resolution of conversion.

(3) *Successive approximations*. Digital values are developed by comparison of an unknown value input signal with a precisely generated reference signal. The operation is like weighing with a beam balance, adding weights to the pans (see Figure 3-21). Successive approximations to the input voltage are made by a D/A converter, which converts the output of the A/D converter back to an analog value. This value is compared to the input analog signal by a comparator circuit. Whenever the input signal is higher than the converter

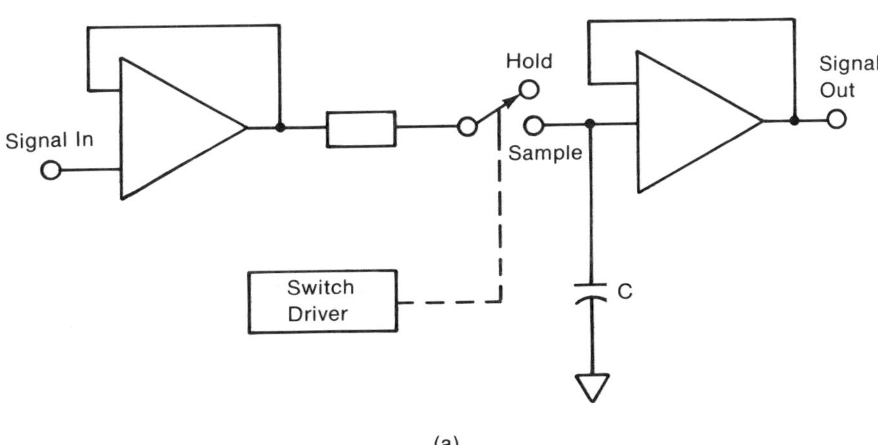

(a)

Sample and Hold — Analog Circuit

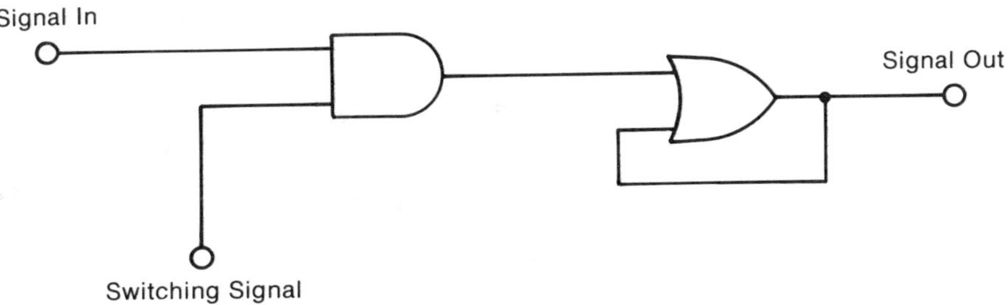

(b)

Sample and Hold — Digital Circuit

Figure 3-19. Sample and Hold Circuits

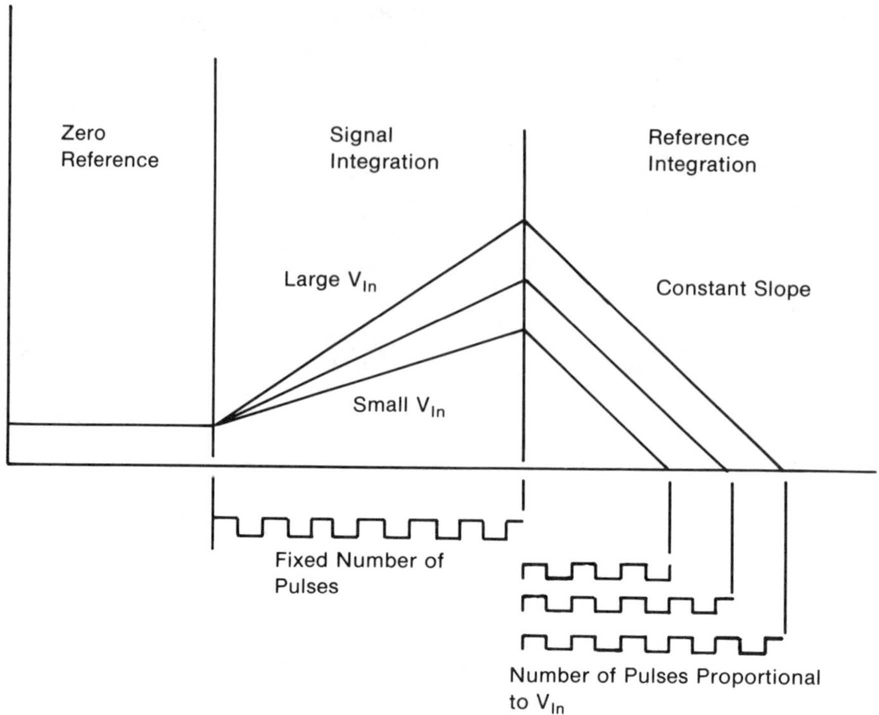

Figure 3-20. Dual Slope Method of Analog-to-Digital Conversion

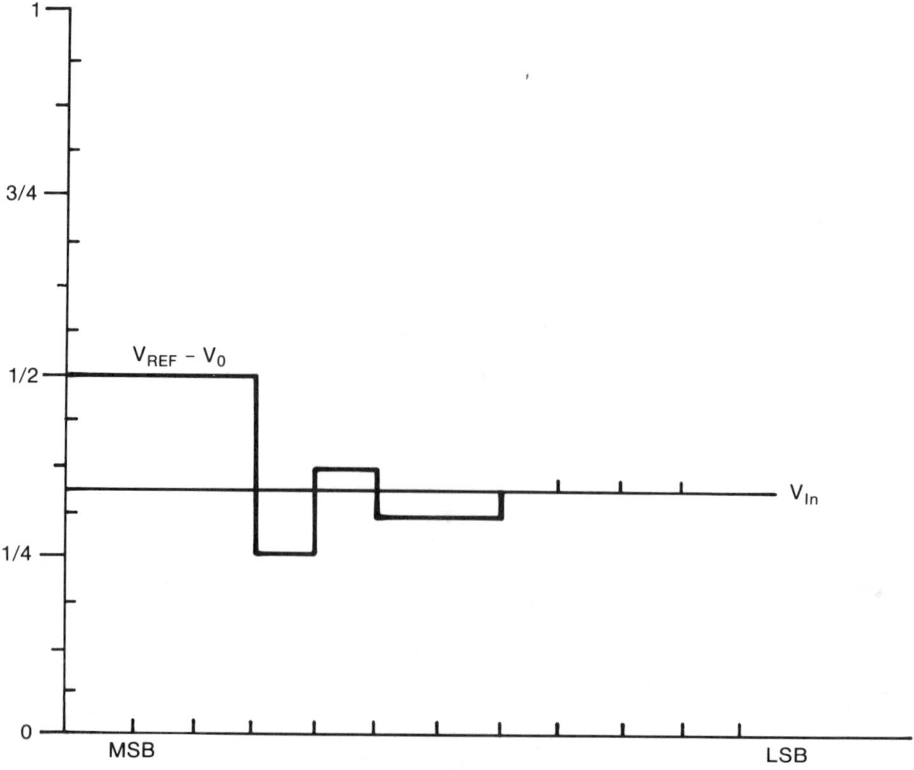

Figure 3-21. Successive Approximations Method of Analog-to-Digital Conversion

output, the comparator output will be true. Starting with the most significant bit, a 1 is tried in each bit position and kept, if the signal is greater than the resulting D/A converter output, in a register addressed for that purpose. The first test is at 50%. If the measured signal is higher, the next test will increase half the amount of the preceding one, to 75%. However, if it is lower, the next test will go down to 25%. The third test will increment by ⅛th, the fourth by 1/16th, and so on.

Since the input signal may change during the conversion period, it must be sampled and then held in a latched register until the conversion is completed. The binary code remaining in the addressed register digitally represents the analog input voltage.

The successive approximation method is very fast. One hundred thousand samples per second can be read. This makes it a good method to use for fast-changing variables such as shock or vibration. On the other hand, since its sampling rate is fast and it has no time to average it out, it must use filtering to cut out noise. Therefore, it has poorer sensitivity for low level signals than some other methods of analog-to-digital conversion.

3-22 Digital to Analog Converters

Output signals that have been developed digitally must be converted to analog values before they can be used by the final analog receiving device. Digital signals, developed at the output of a A/D converter, may be converted back to analog values that can be compared to an analog reference as was described for the successive approximation A/D conversion technique. Figure 3-22 illustrates a method of D/A conversion as it is commonly implemented using resistors and an external operational amplifiers. The conversion is obtained by the appropriate choice of summing resistors and the use of a reference voltage as the analog input. The eight bits of digital input are applied to the switch as drivers.

An equivalent integrated circuit that performs the same sort of function is available as a self-contained monolithic chip. Each input bit switches a current source with the proper binary weight. The currents are summed and converted to a proportional output voltage.

From this stage, circuitry becomes more and more specialized and sophisticated. Many other types of information transfer mentioned in the following sections are available on chips, medium or large-scale integrated circuits, or even entire printed circuit cards. Circuits for handshaking between receiver and transmitter, protocol generation, complete input/ output systems, and many others are used so generally that they have been standardized as chips or integrated circuits. Broken down into their simplest elements, however, they all resolve into switches that store or compare bits of information.

3-23 Programming — Machine Language

All devices and systems made by assembling and interconnecting pieces of equipment have tangible parts, parts that can be seen and felt; and something intangible, the design that defines what the performance should be.

The tangible parts are called hardware for most devices and systems. For mechanical equipment, the intangible part is supplied by operating instructions and assembly drawings. Schematic wiring diagrams supply the intangible design concept for electrical systems. Instrumentation systems use flow sheets, wiring schematics, and process and instrumentation drawings to define how operation should take place. In the vocabulary of computer users the word "software" has become accepted to describe the concepts of operation, as well as the expression of it.

Originally, computers achieved performance by connecting hardware together so that operation took place in a desired sequence of activities. In the digital control devices used today, the sequence of operations that was previously determined by physical interconnections is accomplished by an array of bits stored in ROM chips. The configuration of the bits and the sequence in which they are used is called a program. Since a program accomplishes the same functions as hardware, but without using any tangible devices at all, it is often called "software".

A program, in its most elementary concept, is a sequence of instructions for moving information from one register to another register and for performing binary addition of numbers in registers. The information that is the source of the move and add operations is the state of a bit, or of a number of bits. The command to execute a WRITE action, for example, is given by making the voltage "TRUE" at a specific pin of a specific chip at a specific time in the sequence of operations. This might be done by sending a 1 bit over a bus to the chip. If this action in turn sets "TRUE" values to all the latches of a specific register in the chip to accept inputs and put them in place of whatever values were latched in previously, then the chip will accept values "written" to it on the input bus. The connection to the foil path by specific pins of a chip is the same as hardwiring, whereby the chip can recognize specific binary values as commands and perform as directed by a specific pattern of bits.

These actions are part of the basic operating functions of the computer itself. They are as basic as the functions in the human body that control breathing and swallowing and the passage of food through the digestive system. They take place transparently, but without them nothing would ever happen. There is a large and complicated system of them in a

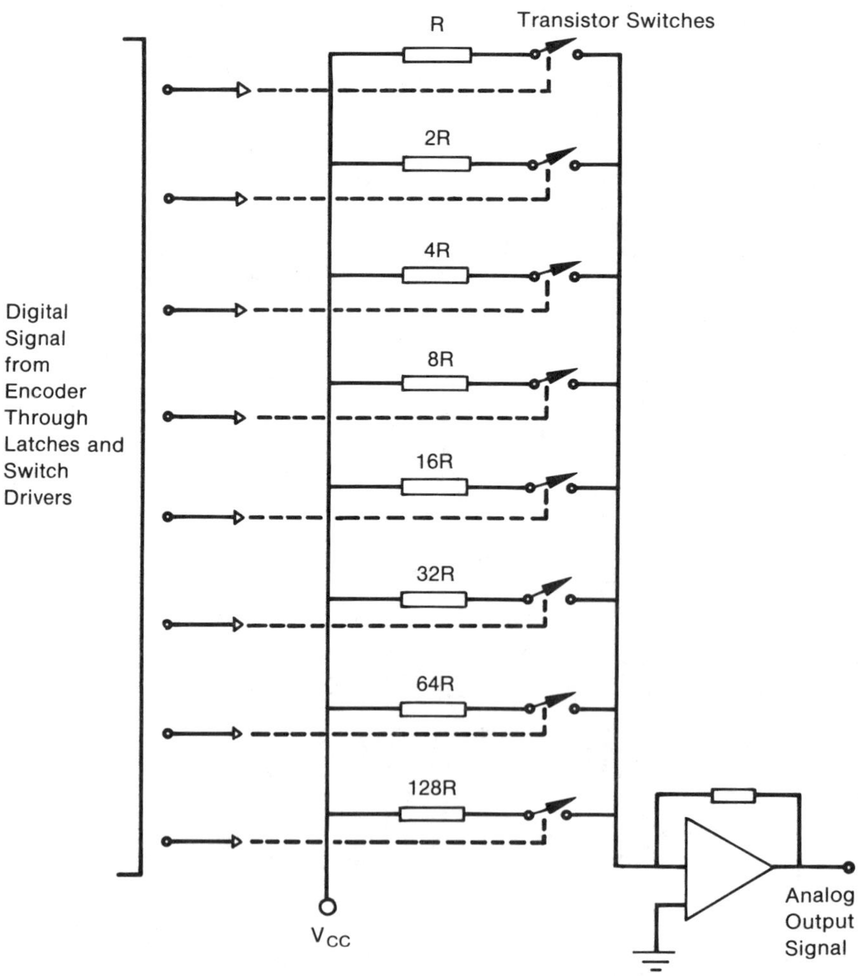

Figure 3-22. Digital-to-Analog Converter

computer, and designers have figured out what they have to be for each particular machine. These basic programs, written in 1's and 0's, are called machine language. It is the only language the computer acts on without the help of a translator.

3-24 Assembly Language

Machine language is time consuming to write, hard to learn, and mistakes can easily be made in selecting the correct arrangements of 1's and 0's. Consequently, application programmers use languages that are really codes for the actions that machine language performs. The next level up from machine language is a coded language that can be translated by the computer itself (using another program resident in its memory) into a series of words that have the bit patterns necessary to perform the machine language operations. While machine language functions at the bit level, assembly language functions at the word level.

As an example of the way assembly language is used, suppose that there are three boxes, one marked A, another marked B, and the third marked C. Suppose also that there is a piece of paper in each box, blank in boxes A and B, but having a word written on the paper in C. Suppose also that it is not permitted to move information from box C to box B, but that information can be transferred between boxes A and B, and between boxes A and C.

If a person were commanded to write onto the paper in box B the word that is in box C, a series of actions would have to take place in order to carry out the command. A very primary set of actions consisting of muscle contractions and extensions would be completely transparent. Without really thinking about it, the individual commanded would extend his hand, open box C, read the message, close box C, open box A (remember, he can't go directly to B), write the message on the paper in box A, read it to himself, erase it from the paper in box A, open box B, and write it on the paper in box B.

Two commands in assembly language would generate all this activity. The first would instruct the computer to repeat the information in box C, putting it into box A but not erasing it from box C. Depending on the microprocessor doing the work, the command might be written

```
MOV [C],A
```

The second command would instruct the computer to repeat the information in box A, putting it into box B, but erasing it from box A. This command might be written

```
MOV A,B
```

When the programmer enters these letters from the keyboard, another program resident in the computer's memory converts the symbolic names for operations and locations into the actual binary machine code that is required by the computer. The resident program is called an assembler. In a computer whose operating system uses 16-bit words, the machine code would be expressed in hexadecimal. The assembler generates a series of machine language instructions that result in specific registers being filled with bit patterns that correspond to hexadecimal words like FE, 7A, 06, and C5. The bit patterns, whatever they may be, will be the ones that are going to do the job called for by

```
MOV A,B
```

and the other instructions in the program. This is represented by the the bit patterns of the hexadecimal words that contain the entire series of machine language instructions that repeat a byte of information from a memory register called C (defined in another part of the program), to a memory register called A, then from that register to another one called B, adding a 0 to each location in A to leave it empty of information.

Assembly language has many commands. They are written in a code that may seem to be obscure, but it is a code that an experienced programmer can use. Assembly language instructions operate very rapidly and use less memory than higher level languages. The fastest operating software furnished with a computer or on a purchased disk is written in the assembly language that corresponds to the microprocessor used by the computer.

3-25 High Level Languages

Instead of using assembly language, most application programmers will use a high level language that is coded at a level where commands, instead of expressing words, correspond to complete actions. Commands and statements making up a program are converted into machine language by other programs called compilers.

There are two kinds of high level language. One kind includes the high level languages for which compilers are commonly available, such as BASIC, FORTRAN, and PASCAL. Table 3-4 summarizes the characteristics of commonly used high level languages. The other kind of high level language can be classified as "high level process control" languages. It is sometimes called "process-oriented" language.

A high level language that permits problem-solving tasks to be carried out through a dialog between the user and the computer is said to be "interactive". One which requires placing certain parts of the program in specific locations, defining variables before they are used, and writing the program in modules that build on the previously defined steps is said to be "structured". Another type of high level language is described as "threaded". A threaded

Table 3-4
Characteristics of Popular High-Level Languages

Language	Characteristics	Relative Difficulty
APL (A Programming Language)	Interactive. Allows a programmer to perform very complex functions with few instructions.	Difficult
BASIC (Beginners All-purpose Symbolic Instruction Code)	Unstructured, interactive. Popular, because it uses English-like commands.	Easy to learn, but difficult if used to maximum potential
C	Somewhat structured.	Medium
COBOL (Common Business Oriented Language)	Limited mathematical capabilities, but well suited for business activities. Not used for digital control applications.	Easy
Forth™	Threaded, interactive, very detailed. Used for robotics applications.	Difficult
FORTRAN (FORmula TRANslator)	Widely used for engineering computations.	Medium
LISP™ (LISt Processor)	Used for artificial intelligence programming.	Difficult
Pascal	Highly structured. Non-interactive.	Difficult
Modulo 2™	Structured, related to Pascal. Not used for process control.	Difficult
Ada™	Structured, related to Pascal. Used by the Department of Defense.	Difficult

language builds a program by rearranging and restructuring instructions to define new commands. The newest development is "artificial intelligence," a form of programming that appears to allow the computer to make intelligent decisions on the basis of an extensive storage of operating history and process characteristics.

There are benefits to programmers in using high level languages. A high level language is a sequence of statements. This makes larger programs easier to understand. High level languages are modular. Subroutines can be written and used over and over again, simplifying and organizing the programming process. They deal with variable names, not addresses. This facilitates branching and jumping and the use of subroutines. It is easier to write "Start Sump Pump" than to look up and reference address 10075. There is an equal advantage when reading, checking, or troubleshooting a program.

In BASIC, one command would execute the example given above.

$$B = C$$

would be read as "The contents of register B is defined to be the value of the contents of register C". The BASIC compiler, when it comes to the command B=C creates the machine code sequence that moves C to A, then to B. Much interpreting and translating must be done, so it takes longer to do it this way. Programs used for sequential operations, where a set of conditions must be initiated *when* something happens and continue *while* or *until* other constraints are satisfied, are often written in Basic.

High level languages such as FORTRAN have a use in applications that perform a lot of mathematics or that need custom algorithms. Compilers for these languages are usually found only in more expensive systems, such as a personal computer or in a minicomputer, where they can be used for many more programming tasks than just process control.

3-26 Specialized High Level Languages

Minicomputers for process control applications can be programmed in almost any standard high level language desired. They do require special programming skills to use them effectively, and at the time of this writing there are not many individuals who combine such a degree of skill with an equal understanding of process control application. Consequently, other categories of digital control devices are often supplied with compilers that are specific to languages modified from standard high level forms by their suppliers so that they can perform specialized process control-oriented functions. These special-purpose languages are interactive, and easier to use than standard high level languages, because most of the programming has been done and is transparent. This makes them more conversational from the user's point of view. Specialized programming skills are not required, although time must be spent to become familiar with the methods and to learn to use them proficiently. No one will be able to sit down at the keyboard and "in a few minutes" be able to configure a system, no matter what promises the advertisements make.

These specialized programming languages have instructions dedicated to a specific application. The "fill in the blanks" routines used by many distributed control systems, the ladder diagram language used by programmable logic controllers, and the plotters used for creating graphics displays are examples of special-purpose languages. They are structured so that the user can think in terms that let the language simulate a procedure with which he is familiar. Languages for programming programmable logic controllers, for example, have commands that duplicate the steps an electrical draftsman uses to make a schematic wiring diagram. Alternately, they may be appropriate to executing Boolean logic and handling discrete inputs and outputs. Languages for drawing graphic pictures have instructions that move lines and shapes and sound like the description of instructions for making a drawing. They may use the words "Circle, 4" to draw a 4-inch diameter circle, or they may accomplish the same thing by touching a light pen to a picture of a circle. Filling in the blanks in a distributed control system configuration display puts subroutines with names descriptive of the service they perform, stored in ROM and containing commands to perform specific process functions, into specific memory locations that are defined by the cursor position on the screen. The subroutines are called algorithms. There will be an algorithm for summing, for example, and the programmer will fill into the display, from the keyboard, a configuration code that moves the summer subroutine into a portion of Random Access Memory where other codes identify registers that hold input value information, gain values, and all the other information the algorithm needs to do its job. This area of RAM is like a scratchpad where a design engineer makes sketches.

Requiring even less memorizing of codes is a language, beginning to appear in Japanese-designed systems, that allows the process flow sheet to be drawn into a graphic display on the screen. This is all the software needs to develop the configuration of the algorithms and complete the programming.

3-27 Program Storage

The interface between hardware and software is the memory in which the programs are stored. The first basic commands of a microprocessor operating program are stored in its read only memory (ROM) and are activated when the power is first turned on. These may only define where to go for the first instruction, and which register to put it into, but that is enough to get things going. This bootstrapping action is the source of the expression "booting up".

The balance of the computer's operating program will be stored in a permanent memory, either in chips or on tape or disk. One of the first operations after loading the computer will be to read this into a working area in RAM. Sometimes the chips holding instructions for subroutines that are used over and over again in the operating program are in a type of ROM called Programmable Read Only Memory (PROM). A PROM is read into only once. It is a convenient way to transport a program; the functions of the computer can be changed by plugging in new PROMs. Another type of memory chip is called EPROM. This is Erasable Programmable Read Only Memory, which can be cleared of information by radiation from fluorescent light so that the program can be changed by downloading new information into it. The new program in its machine language form is downloaded — transferred over a communication link one bit at a time — from the memory area where it is stored, into specific ROM registers. This is a useful way to keep memory up to date if there are a number of revision levels.

Minicomputer application programs are not permanently stored in the computer memory chips. They are more likely to be stored on tape or disk, from which they are loaded into a working area in RAM. Process-oriented information that has been entered into memory for storage will be in RAM, which is volatile. Information in this type of memory will be lost if intrument power is lost. To maintain variable data such as set points, tuning constants, and current values RAM should be powered from long-life batteries if the normal source of power becomes unavailable.

REFERENCES

1.   Jury, Eliahu I., *Sampled Data Control Systems*, New York: Wiley and Sons, Inc. (1958).
2.   Anonymous, *Digital Techniques*, Heathkit/Zenith Educational Systems, Prentice-Hall, Inc. (1983).
3.   Cassell, Douglas A., *Microcomputers and Modern Control Engineering*, Reston Publishing Co., Inc. (1983).
4.   Benedick, Dale, "How to Keep Input Signals 'Clean'," *I&CS*, March, 1985, pp 29-32.

5. Malmstadt, Howard V., Christie G. Enke, and Stanley R. Crouch, *Electronics and Instrumentation for Scientists*, The Benjamin/Cummings Publishing Company, Inc. (1981).

EXERCISES

1. What is bubble memory? Where is it used in digital control devices?

2. What is the difference between a compiler and an interpreter?

   Between source code and object code?

3. Explain how doping changes a poor conductor into a semiconductor.

4. Explain the relationship between timing and message transfer in an assembly language command execution.

5. Give an example of the way that larger components are made up of switches; and how machine language programming operates the switches to execute commands.

UNIT 4
Systems for Digital Control

Just as beauty is in the eye of the beholder, the meaning of the word "system" is in the context of the user. The word will be used frequently in this text, and it is, therefore, important to define it before using it. For the purpose of the text a system is a set of devices working together to change material and energy, with the objective of producing something. Usually, but not necessarily, the product is tangible. It might, however, be as esoteric as a mathematical computation, or a feeling of pleasure. An industrial process is a system of equipment and instruments that produces a product from raw or partly finished materials. An automobile is a system of electrical, mechanical and hydraulic subsystems. There are always inputs to and outputs from a system. An energy balance and a material balance can always be made around a system. While there may be changes to material and energy in a system, neither is ever created. In totally electronic device systems, there is normally no change in material, only of energy.

In the context of this section two kinds of systems will be considered. First, the digital control device will be presented as a system, a small specialized system made up of the subcomponents outlined in the previous section. Then the larger process control system, making use of smaller, specialized digital control devices, will be examined. The larger system will include some kind of manufacturing process. It will include input devices used to convert process variable measurements to analog electrical signals; the digital control devices themselves; peripheral electronic devices including printers, auxiliary computers, and the network of communications connecting them together; and regulating devices that convert the electrical signal outputs from the principal digital control component into physical outputs to produce process conditions, machine operations, records, or alarms.

As an example of the difference between the two kinds of systems, a single-loop controller is a "system" in the small system category. It is made up of components that perform input and output functions, other components that make the analog-to-digital and the digital-to-analog conversions, memory, and buses for transferring information between components. In a large system, the same single-loop controller might be a component used for feedback control, regulating temperature by controlling the flow of steam through a valve to a heat exchanger in accordance with a measurement of the process temperature.

4-1 The Digital Control Device as a System

Movement of information as quickly and as efficiently as possible is the governing objective in the design of every piece of digital equipment. There is a flow of information through every system of digital equipment: into the system from measured process values, bulk storage, and peripheral computing devices; within the system through I/O devices, signal conditioning, data storage, and logical operations; out of the system to final control elements, recording devices, and peripheral computers.

For a simple digital device such as a pocket calculator, Figure 4-1(a), inputs are entered by finger contact with the keyboard; information is output to a liquid crystal display. Digital logic operations occur in the device itself (Figure 4-1(b)).

The flow pattern is established, even in this relatively unsophisticated device. The keyboard system interfaces with memory and produces bit manipulation; the bit manipulation is associated with information transfer between memory, input/output, and logical operations; the results of the logical operations are made available to the user through operation of a display. The entire activity is energized by regulated power supplies. It is rigidly controlled on a time basis, under the direction of a program.

4-2 Packaging the Electronics System

Putting together digital subcomponents described in the preceding unit can produce an infinite number of complete electronic products, each performing a separate function. Broad categories include data acquisition units, minicomputers, controllers, word processors, etc. Any unit in these categories fits our definition of a small digital system. Each of these digital systems includes circuitry to move information into and out of the unit, perform signal conditioning and logical operations, and store data.

In Figure 4-2 subcomponents assembled into a small system make up a hypothetical data acquisition system. A microprocessor controls the entire operation, delegating specific activities to subordinate microprocessors. This distribution of responsibility leaves the principal processor free to schedule activities, run background diagnostics, and handle interrupts.

The electronics module for data acquisition systems will almost always include an input multiplexer, a programmable gain amplifier, circuitry for signal conditioning, linearizing, and filtering, a sample and hold circuit, an analog-to-digital converter, timing and control logic, and buffers for holding output signals. The systems may exist in the form of a number of chips mounted on a printed circuit card with buses made of foil and flat ribbon connectors connecting the various devices, or it may be built into one large scale integrated circuit chip.

On the printed circuit card, information is moved between registers and components over buses that provide separate paths for communication between processing subcomponents and memory and for input and output. Programming accomplishes the designated objectives of the equipment. The system assembled as shown in Figure 4-2 is generic. The same components are used for minicomputers, distributed control systems, and single-loop controllers. If the name of the device had been changed, the picture could still have had a similar appearance. Objectives of design and combinations of equipment, of course, vary infinitely, but the basic parts are common to all digital control devices.

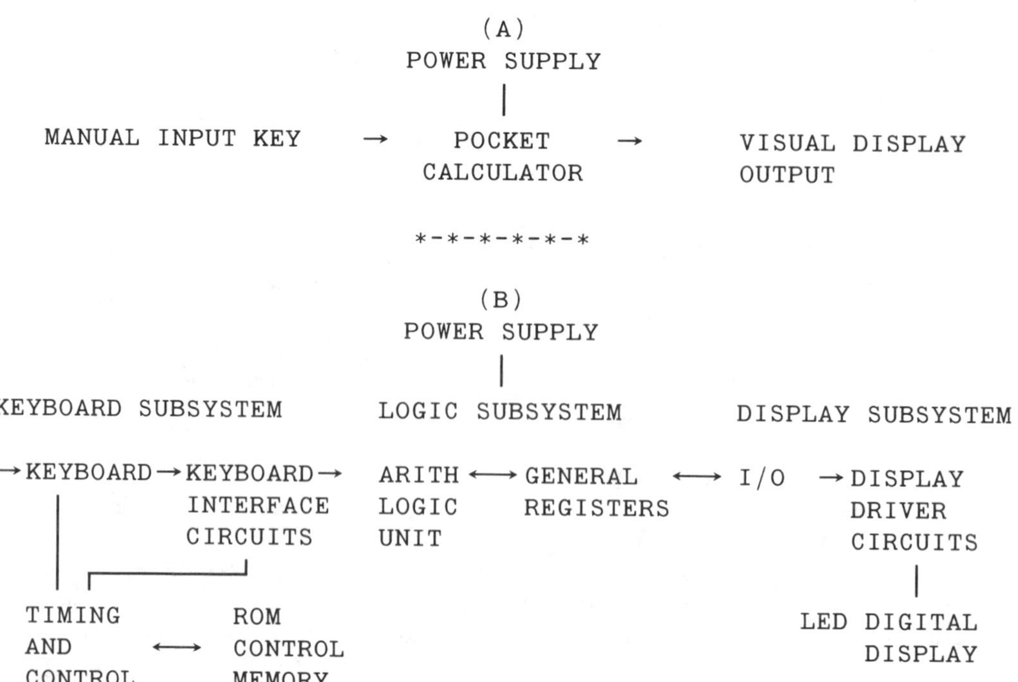

Figure 4-1. Information Flow Through a Digital Control System

In the world of solid state electronics a single chip can contain an entire system. More generally, however, the basic hardware unit of the electronics system package is the printed circuit card, a laminated plastic card on which conductive paths (called buses) of foil are laid. Circuit elements (chips, resistors, switches, etc.) are mounted on the card or in sockets. In either case, pins are in contact with the foil, usually held in place by soldering. Clocks and power supplies will probably be mounted separately. On the other hand a single board may include a real-time clock and regulated power supplies, as well as RAM and ROM memory chips.

Packaging on printed circuit cards promotes standardization. The cards in an assembly are usually of one standard size and shape. A multi-pin connector at one edge allows each card to be plugged into a socket. If more than one card is used in a system the supporting structure for the sockets may be another printed circuit card with foil buses that interconnect a number of sockets into which individual cards can be plugged. This design is called a backplane. An assembly made up of a number of cards plugged into a backplane is called a card file. One or more card files may be mounted in a housing, usually a metal cabinet, and interwired to create a product.

A stand-alone component (packaged into a separate housing) that requires programming will have a built-in keyboard as part of its front panel. This will have keys that can be used to program ranges, arithmetic functions, and sample timing.

4-3 The Complete Digital Control System

The electronics device illustrated above is only one part of the complete digital control system. A full complement of subsystems that could be included is listed in Table 4-1.

4-4 Power Supplies

Well regulated power must be supplied with, or be made available for, each digital control device. The source will usually be industrial plant ac power, and this must be converted to dc of a suitable level and stability. Several levels of power reduction and regulation may be necessary in a system. 24 volts dc is a commonly used primary direct current level, while ±15, ±12, and ±5 volt dc are frequently used for secondary levels.

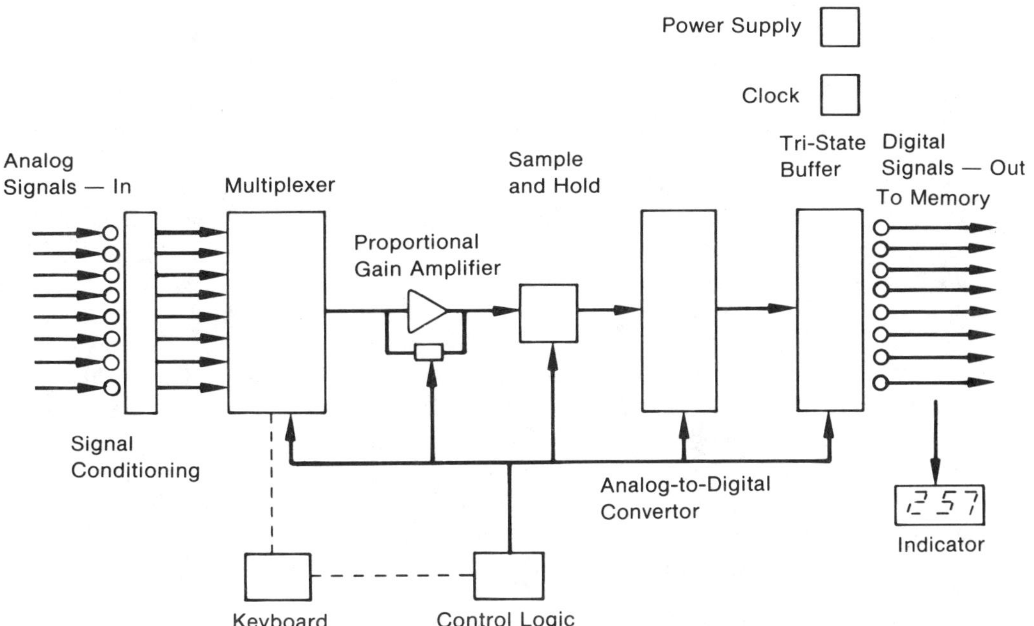

Figure 4-2. A System of Digital Components

Table 4-1
Subsystems Making up a Complete Digital Control System

1. Power supplies
2. Discrete measuring devices
3. Process variable measuring devices
4. Signal conditioning equipment
5. The electronic equipment itself
6. Wiring connecting inputs and outputs to the electronics equipment
7. Final process regulating devices
8. Peripheral equipment to the electronic device
    A. Disk drives
    B. Keyboards
    C. Monitors
    D. Printers
    E. Minicomputers
9. Backup power

The first stage will usually be a supply that is large enough to take care of the entire system, is not extremely highly regulated, but is stable in the presence of ac sources that have transients or are variably loaded. Supplies included for conversion and regulation of power at this stage are designed so that the output will be steady over 0 to 100% of load change even though the input ac varies. Limits of ±5% to ±15% are usually acceptable. A typical specification for the main power supply of a large digital control system might be for an output variation of ±2 volts, around a nominal value of 24 V dc for an input variation of ±15 volts around a nominal 115 volts ac. This will be related to a temperature specification, to limits of weight and size, and also to audio and radio frequency noise.

To get better regulation for specific requirements, this source can be reduced and regulated. Where regulation of dc power for individual circuits must be in the order of a fraction of a volt, separate, highly regulated supplies will be included for the various sub-components that require such regulation. While the digital control device built onto a single printed circuit card may have a small power supply mounted on it, in order to supply the specific levels of voltages its subcomponents need, it is more likely to depend on an externally mounted supply. For example, a group of signal conditioning cards, plugging into a backplane that is part of a 19-inch rack assembly, may become a working module when another card that contains a regulated power supply is plugged into the same backplane. Power supplies will be an integral part of self-contained data acquisition instruments, single-loop controllers, and minicomputers. For programmable logic control and distributed control, the power supply is usually a separate component, mounted in its own housing.

Two types of power supplies are commonly used for digital control equipment. The simplest but less efficient of the two is the series pass regulator, sometimes called a linear supply. This uses a transformer to reduce ac supply voltage (in the United States this is usually 120 volt, 60 Hz) to several lower levels, which are then rectified. The resultant dc voltage is filtered. Often a step-down transformer will follow the rectifier. A more efficient power supply, called a switching supply, uses a low-impedance transistor switch to open and close periodically between input and output. The input alternating current is rectified and filtered into dc with a small amount of ripple. Transistor switches chop this dc voltage into a series of rectangular wave forms, which are transformed to the required voltage level by either an inductor or a transformer. The final output is obtained by rectifying and filtering the voltage adjusted signal.

Much of the circuitry for switching can be built into integrated circuits, taking up only a fraction of the size required by the filters and transformers of linear supplies. Power dissipation and heat generation are low. This results in a significant increase in efficiency over that of the linear supply. Ninety percent efficiencies are claimed, compared to 75% efficiencies for linear supplies.

An additional advantage that switching supplies have is the ability to sustain output for a short period of time — 20 to 40 milliseconds — if there is a total loss of input power. Using diagnostics to detect a failure condition, volatile data can be read and stored in permanent memory during this brief period before total loss of output so that backup units can continue operation of critical processes.

Devices auxiliary to the power supply, usually not supplied by the digital control device producer, should always be considered by the system designer to assure that consistently reliable power is available. Electronic devices cannot operate if supply power is too high or too low. Usually a 10% to 15% variation is allowable. If the available ac power will vary outside published limits because of plant load variances, such as startup and shutdown of heavy equipment or reduction of utility supplied power, it is recommended that a constant voltage transformer be installed. A constant voltage transformer compensates for changes in input supply to maintain a constant output voltage.

Another type of disturbance harmful to the operation of all digital control devices is that caused by electromagnetic interference from high frequency equipment such as welding machines and the field of large motors. If this situation occurs, consideration should be given to installing an isolation transformer ahead of the digital control equipment.

Surges in power resulting from nearby lightning strikes can induce transient voltages and cause a system to fail. The use of commercially available lightning suppressors is advisable for protection of systems installed in locations where severe electrical storms are prevalent.

4-5 Discrete Measuring Devices

Information flows in the digital control system from sensors to the digital control devices, and from them to final control elements.

The on-off signals used by logical control systems may represent an operational condition (yes, the circuit is closed; no, it is not closed) or a process variable limit. In the first category, the most common signal sources are relay contacts and manually positioned push buttons and switches. Examples of process variable limit sensors are bimetallic switches for temperature, floats for level, and proximity detectors or lever-actuated switches for position.

Figure 4-3 illustrates typical discrete input and output signal interfaces. In each case, optical isolation blocks electrical noise in the external circuitry, preventing it from interfering with the functioning of the integrated circuits to which the signal is directed. One way to achieve optical isolation is to use the signal to light a light-emitting diode (LED), then use the light to activate a light-sensitive transistor.

For the input circuit, Figure 4.3(a), a switch closure connects 120 volts ac to the input terminals through a pair of twisted, shielded wires. The voltage is rectified, filtered, and conditioned for use in the computer's TTL circuit, as an indication that the switch is closed. The output interface converts a TTL signal to an ac voltage which fires a triac to develop power sufficient to operate a relay or solenoid valve. (See Section 4-10).

4-6 Process Variable Measuring Devices

The 1980's must be considered a time of transition from sensors that transmit continuous analog signals to ones that develop digital outputs. The automotive industry is developing sensors, primarily for pressure and vacuum measurement, with digital outputs. These are for use in the energy management systems in automobiles and trucks, and the results of their successful use will probably appear in commercially available process measurement devices in the future.

Some digital transmitters are already being used for specialized applications. A level measurement system for tank farms, described as an application example in Unit 12 of this text, transmits information in digital form over a data highway. Another system for measuring gas composition uses digital transmitters that are battery-powered and connected back to their electronics package receiver by fiber optics cable. Each unit includes its own microprocessor, and so each unit becomes a special-purpose computer. Using batteries eliminates running power wiring and is a step toward intrinsic safety. Digital circuits have very low current flow and therefore use little power. Using fiber optics eliminates the interference that electrical circuits in metallic conductor wires are subject to: electromotive interference, cross talk, and induced voltages from transients. Optical multiplexers can be built into software so that only one input at a time is processed.

Honeywell's ST3000™ digital differential pressure transmitter has a microprocessor that allows the transmitter to be calibrated from a remote location, using digital signals sent from a hand-held device. This transmitter (nicknamed the "smart" transmitter) makes a temperature and a static pressure measurement in the same housing with the dP cell. The three signals are combined with calibration data stored in memory. The output is a linearized, extremely accurate differential pressure measurement, compensated for static pressure and temperature.

For the present discussion it will be assumed that process variable inputs come from analog-type transmitters. The most commonly measured process variables are temperature,

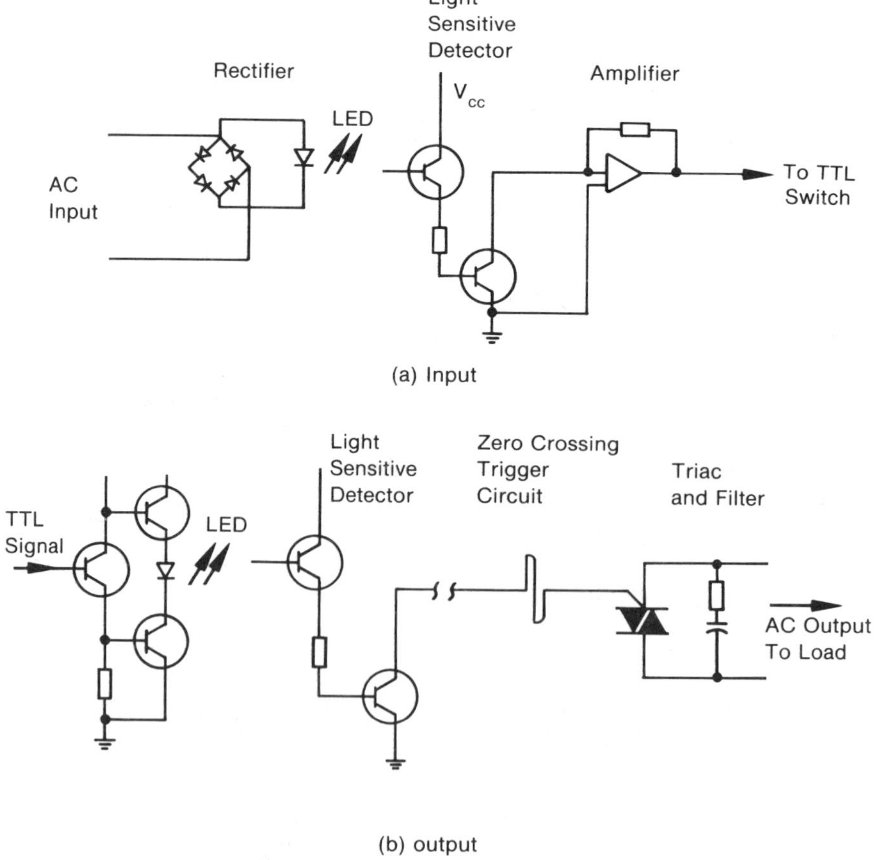

Figure 4-3. Discrete Control Interfaces

pressure, flow, and level, used to infer the changes that are taking place in the process. Specially built transmitters may exist to take real-time measurements of composition, energy content, or percentage completion, but commercially available devices for this type of measurement are either not developed or are very expensive. In many cases, fortunately, these very important variables can be inferred from measurements that can be made, i. e., temperature, etc. The balance of process measurements come from a large selection of variables, including weight, moisture content, humidity, vibration, viscosity, gas analysis, and many, many more.

The relationships between variables and signal types are summarized in Table 4-2.

Table 4-2
Analog Signals Associated with Commonly Measured Process Variables

Signal Type	Process Variable	Sensor
Millivoltage	Temperature	Thermocouple Resistance thermometer Radiation detector
Current	Flow, level, differential pressure, relative humidity, and most other variables	Two-wire transmitter (converts low-level signal from strain gage, etc., to current — 4 to 20 mA is most common)
Pulse	Flow	Turbine, magnetic and positive displacement meters
	Speed	Tachometers
High-level voltage	Linear position	LVDT (Linear Variable Displacement Transmitter)
	Rotary position	Potentiometer

4-7 Signal Conditioning, External to the Digital Control Device

If the digital control device, whichever of the types listed in Table 4-2 it might be, is designed to receive the transmitted process variable signal directly, no additional equipment need be supplied. Many digital control devices are able to do this, but not all commercially listed devices can handle every type of input. Consequently, there are analog-type components, designed for external (usually relay rack) mounting, that can be used to accomplish the necessary conversion to 4-20 mA, along with linearization, filtering, and span and bias adjustment.

Linearization may be required because of impurities in the material from which sensors are made; because of mechanical stresses in the components of a sensor; because of mechanical nonlinearities introduced by linkages; or because the output of a sensor measuring a process variable is influenced by the energy required by the sensor for making the measurement. For example, thermocouples are used to measure temperature. A thermocouple is a junction of two different metals across which a current flows when the temperature changes. The change of current in relation to the change in temperature that produces the signal may not be linear at every point over the temperature span because of impurities in the metals that form the junction. The nonlinearity is predictable, however, if standard compositions are used, and the signal can be electrically conditioned — changed so that the input to the computing circuit has a consistent change in value for every change in temperature.

The two most popular inputs to digital control devices are 0 to 10 volts dc (used by many programmable logic controllers) and 1 to 5 volts dc (used for analog variables by data acquisition, distributed control, single-loop control, and minicomputers). Signals are often produced in the form of milliamperes. One to 5 volts is readily obtained by passing a 4 to 20 mA current input to signal common through a 250-ohm resistor.

In industrial process control systems many sensors develop pneumatic signals. Three to 15 psi and 6 to 30 psi are common ranges. In order to be used by a digital control device the pneumatic signal will be changed to 4 to 20 mA. Devices designed to do this are of such a size that groups of 12 to 16 of them will mount on a 19-inch plate for assembly into a relay rack containing the electronic equipment.

Sometimes, to reduce the cost of transducers, pneumatic signals are multiplexed. A scanning device refers each pressure consecutively to a strain gage transducer. As many as sixty-four 3 to 15 psi signals can be multiplexed to one transducer, which develops a 4 to 20 mA signal. When the scanning device is used with a computer, a stepping motor controlled by the computer drives a pneumatic scanner and a photodiode encoder. This produces a binary, BCD, or decimal code to identify each measurement. There is a very small volume between the scanner head and the transducer, and rates of 6 inputs/minute are possible. Faster rates can be obtained by connecting one signal to several input ports. The effect of this additional sampling time constraint must be evaluated in the context of process response if it is to be used for control.

A digital interface has been developed for serial or parallel transmission to a computer of the scanner's digital output signal. This device includes a keyboard for programming scan rates and range data, runs self-diagnostics, and indicates input values on an LED display. It is an example of a peripheral digital control device.

Differences in electrical ground location between signal site and measuring circuit complicates interface design problems. Amplifiers receiving input signals must have good common mode rejection capability, high input impedance, low drift, and adjustable gain. An isolation amplifier may be required to accommodate very high common mode voltage levels or very low common mode leakage current requirement. Optical isolation, or transformer-coupled isolation can be used for this purpose.

Voltage inputs must be isolated so that each scanned value is unique. One method is to use a flying capacitor, illustrated in the preceding unit (see Figure 3-19). Another is to use a resistor divider to reduce all signals to a millivolt level that can be isolated by a very high impedance amplifier. Still other systems use photo-isolated inputs.

4-8 Connecting Wiring

Interconnections between components carry information and are an important part of any system of digital control equipment. It might be more correct to say "interconnecting signal conductors" because fiber optic cable is becoming seriously considered as a replacement for metallic wire conductors in circumstances where its unique qualities offer a solution to interference and corrosion problems. The hardware aspects of conductors are included in Unit 5 as part of the discussion of communications.

Attention to installation of wiring must be considered in the context of the system. Good practice contributes significantly to satisfactory system performance. Metallic wire conductors must be installed so as to minimize electrical interference. Failure to pay attention to this can contribute to the "garbage in, garbage out" characteristic of digital instrumentation.

Wiring carrying discrete on-off signals is usually not bothered by interference because only the presence of one of two possible voltage levels must be detected. Relatively heavy (No. 12 or No. 14 AWG) wire is used, with no special shielding or twisting.

Wiring carrying analog signals, however, is subject to a number of varieties of interference and more care must be taken to avoid signal errors. The following interferences can occur:

(1) Difference in ground potential at the two ends of a transmission line can be a source of current flow through ground from one point to the other. This "ground loop" introduces false voltage values.

(2) Inductive or capacitive coupling can occur between the transmission line and adjacent ac or dc lines on which switching occurs.

(3) Motors and generators adjacent to transmission lines produce electromagnetic fields that can distort signals.

(4) When circuit breakers close to turn on heavy loads (air conditioners, elevators, and crane motors, for example) ac voltages may drop low enough to affect power supply regulation.

Trying to analyze all the reasons and all the cures is not a function of this text. Special cases will require special study and special corrective measures, such as using isolation transformers or running wire in metal conduit. There are, however, good installation practices that can reduce the likelihood of occurrence of a majority of the problems that can exist, and so they are listed. The recommendations include the following:

(1) AC and dc wiring should be separated by at least eight inches. They should be run in separate wire trays.

(2) Low level wiring, carrying dc at millivoltage levels, is especially vulnerable to picking up adjacent voltages. Thermocouple extension leadwire, for example, should be routed so that it crosses ac lines at 90 degrees.

(3) Shielded twisted pairs should be used for dc wiring that carries analog signals. The currents carried do not justify the size, but, to have adequate pull strength, wire size should not be less than No. 16 AWG.

(4) Connect shields to ground at one end only, preferably the end terminated at the digital control device.

(5) Most digital systems call for one, and only one, system common connection, no matter how many components there are in the system. Each instrument's system common point should be connected separately to one junction point, preferably a substantial bus bar, using wire heavy enough to assure that current flow in it will not exceed a few microamperes. From the single common point, a heavy cable should complete the connection to earth ground.

(6) Suppliers' recommendations for grounding should be studied and followed.

4-9 Final Regulating Devices

If the digital control system has been designed for process control, there will be outputs to regulating devices in the process area.

Signals for analog control may operate any of the devices listed in Table 4-3.

Table 4-3
Final Regulating Devices

1. Diaphragm-operated control valves
2. Reversing drive motors
3. Silicon-controlled rectifiers
4. Stepping motors
5. Solenoid valves and relays

(1) In industrial process control systems, diaphragm control valves are most commonly used. The valve body, like the valve controlling flow of hot and cold water in a domestic kitchen sink, is moved away from or closer to a port to throttle the flow of fluid in a process piping system. It is usually operated pneumatically, its motion controlled by positioning a shaft moved by a flexible diaphragm in proportion to the amount of air pressure applied to it. A typical range of air pressure is 3 to 15 psi. The output for control from the digital control device will normally be an electrical signal in the range of 4 to 20 milliamperes of current. If so, there must be some sort of electrical-to-pneumatic converson. Figure 4-4 illustrates a typical control valve installation, used as the regulating device in a feedback control loop.

(2) Some suppliers provide a different type of control signal output for large dampers and control valves requiring high torque operators. This type of regulator is often used in steel plants and electric power producing utilities. The output signals used are electrical pulses for driving an electric reversing motor. The motor has two windings, one producing clockwise rotation of the motor shaft, the other counterclockwise rotation. Through gearing, the motor rotation positions an output shaft, moving it a fraction of a degree in response to a pulse calling for clockwise or counterclockwise motion. The motion of the output shaft, mechanically coupled to a valve shaft or to a damper, develops the final control action. Pulses to provide the clockwise or counterclockwise rotation can be variable in duration, as well as in frequency, as a function of the amount of control action required by the digital controlling device. In order to stabilize the positioning action, an internal feedback loop is formed from a measurement of the drive unit position. A voltage across a feedback slidewire provides a signal that corresponds to drive unit shaft position. This is fed back and matched to the signal that defines where the drive unit should be. Corrective positioning will continue until the difference between the two values is reduced to zero.

A pneumatic device called a valve positioner performs the same stabilizing action for a pneumatic control valve installation to assure accurate positioning.

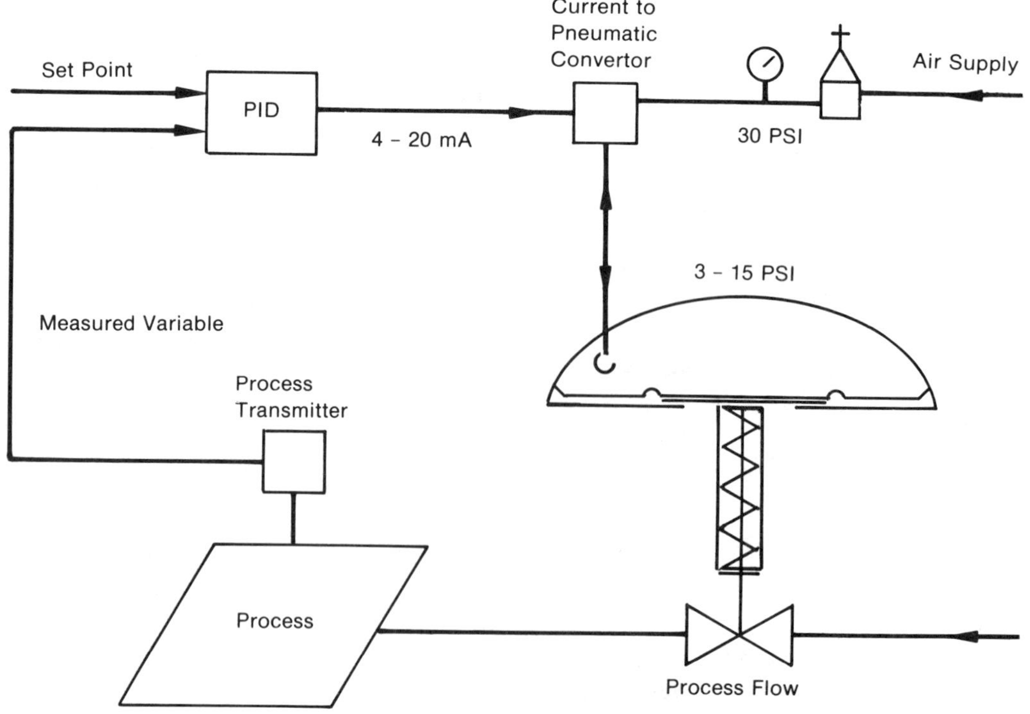

Figure 4-4. Feedback Control System with Control Valve Regulating Flow

(3) Four to 20 milliampere output signals can be used directly to operate resistance heater control circuits. These function to regulate electrical power by switching alternating current power, used for heating elements, on and off for a portion of each cycle. The 4 to 20 milliampere signal from the digital control device is amplified by a power controller to produce a signal that will drive a silicon-controlled rectifier. This is a variable impedance that proportions power to an electrical load, accomplishing the same function as a rheostat.

Large heaters operate from a saturable core reactor, a magnetic device with an iron core and two sets of windings. One set is connected in series with the load, and the inductance of this winding is used to attenuate the load current. This set is called the gate winding. The other winding, called the control winding, is used to vary the flux in the magnetic core and thus vary the impedance of the gate winding. The 4 to 20 mA signal operates through a controller and a silicon-controlled rectifier to bring about the variation in flux.

(4) The 3 to 15 psi signal for valve operation can come from the action of a train of digital pulses. This is used to actuate a stepping motor that positions the air regulator in a 3 to 15 psi pneumatic controller, so that the reference settings in the controller are changed in accordance with the digital control device's computed reference values. The pneumatic controller performs the operating function, the digital control device tells it what to do. This arrangement allows a user to modify and retain existing pneumatic equipment, interfacing a computer to it.

(5) Control outputs can be discrete signals and still perform proportional control. The simplest control signal is an on-off signal to a load. The thermostat in a home hot water heating system puts out a discrete on-off signal to start and stop the hot water circulating pump. Regulating devices operated by on-off signals include control valves in which a plug on the end of a shaft opens and closes a port as the shaft is moved by a solenoid energized in response to an on-off electrical control signal; and relays, with contacts on an armature moving in response to the energization of a coil in response to an on-off signal. The relay contacts take open or closed positions to interrupt or pass currents for operation of electrical equipment.

A variation called duration adjusting control is used for systems with very slow response, such as steel billet heating furnace control. In this case there is a signal to a solenoid valve coil to turn on gas or oil to the heating system. The valve is energized for a period, then de-energized for a period. The total on-off period is adjustable, and the percentage on time is also adjustable. This control form produces less deviation from reference temperature than the more simple on-off thermostat type of control.

4-10 Discrete Output Signals

Signals for digital logic control will be discrete, supplying power for actuating solenoids, relays, and servomotors to control machine tools, robots, and other mechanical equipment. These electrical outputs may be triggered by TTL circuitry that actuates triacs. A triac is a semiconductor that behaves much like two silicon-controlled rectifiers connected back to back, so that it conducts in both directions and can be used with alternating current. When the triac is "turned on" by the actuating signal, it conducts and powers the electrical load to which it is connected. There is always a small amount of leakage current through a triac, and, if the load has a low threshold current (sometimes the case with annunciator circuits), an interposing relay should be used. Figure 4-3(b) illustrates an optically isolated output from TTL logic.

Digitally actuated valves, through operation of solenoids that are energized according to a digital pattern, have been on the market for a number of years. They are not widely used; one must suppose that the price is too high for the benefits of extra resolution. Figures of $10,000 per inch — that is, $30,000 for a three inch valve — have been quoted. Another disadvantage of this type of valve is a large amount of pressure loss; there are many sharp turns in the flow path because the various solenoids have to be connected into a header. However, if ten solenoid valves are used, sized to correspond to powers of 2 (1, 2, 4, 8, 16, 32, 64, 128, 256, 512), a resolution of one part in 1024 is possible. For the process that requires it, cost will not be a factor.

4-11 Peripheral Equipment

Communication between the digital control devices and the equipment located in the process areas is an exchange of voltages and currents. Information flow between these devices and other electronic equipment peripheral to them is formulated in binary notation, often, but not necessarily, in ASCII code. The system will usually have peripheral devices that produce or use this data flow, including keyboards, bulk memory storage, monitors, and printers.

Monitors

A cathode ray tube (CRT) video screen is the standard means of displaying information to interface an operator with the process his digital control equipment is taking measurements from and controlling. Screens will vary is size and display from small (9-in. or 12-in. diagonal) monochrome displays with no graphics capability to larger (19-in. diagonal) multicolor graphic displays. The former are used for programming and for message display; the latter are identified with process control systems.

Some screens include a grid of infrared beams or fine wires to replace some of the keyboard tasks. For these models, touching the screen with a finger causes display in the region touched to be changed. This facilitates menu-driven programs, alarm acknowledgment, and the calling up of details from a larger group. Some systems operate in the same fashion from the placement of a light pen.

The screen of a cathode ray tube functions as a display device through the stimulation of a phosphor coating by a beam of electrons. Three concurrent beams are used for multicolor displays, producing mixtures of red, green, and blue phosphorescence. The beams are moved across the screen, scanning a series of lines from top to bottom and repeating the scan. Programming turns the beams on and off for each of thousands of screen locations. The CRT is a digital computer in its own right; programming creates characters, changes backgrounds, causes blinking, and has, in fact, created a new graphic art form.

Keyboards

Every operating system which includes the capability for an operator to issue instructions to the digital control equipment will include a keyboard, built into the equipment housing or connected to it by a short length of cable. The keyboard is a computer that uses the bit patterns generated when keyboard keys are pressed to send information to the monitor for display, to activate stored routines, to acknowledge alarms, to call up operating routines, and to point to stored information about specific parts of the system. Each keystroke sends a binary message that may be interpreted directly or point to a section of memory where entire routines of instruction are stored.

There are many variations in the layout of keys. The standard QWERTYUIOP of the typewriter keyboard is most often used. Sometimes there is a choice between that arrangement and an alternate arrangement attributed to August Dvorak, selectable by changing a jumper or switch position. Greater typing facility and greater word forming power is claimed for this arrangement. With the letters on Dvorak's home row — AOEUIDHTNS — the typist can produce about 3000 common English words. The ASDFGHJKL of the more common home row arrangement makes fewer than 300 common words.

Many keyboard arrangements will include the standard number pad of the accounting calculator keyboard, with a HOME button and four arrows for up, down, left, and right movement of a cursor. In addition, there will be many special-purpose keys and often some that can be programmed to suit the user's specific needs.

Keys may be raised, rounded, flat, or concave. They may be printed on a flexible membrane, a plastic sheet that has become familiar through extensive use for calculators and microwave ovens. They should depress with a satisfying tactile resistance and an audible confirmation of their travel, and sometimes this is built into the membrane switch pad.

Keyboards are usually equipped with a lock that must be opened before certain key-actuated functions are accessible to an operator. This is a safety feature that protects the system from use by unauthorized individuals. Most keyboard use is in conjunction with a video monitor, where the results of key action are displayed. One way in which the key lock is made effective is to deactivate the screen cursor when the lock is not in an EDIT position. Without a cursor it is not possible to define the location of information, and the screen can be used for viewing only, not for changing programmed information.

The switches operated by the keys may have mechanically positioned contacts, they may be encapsulated reed switches, or they may be non-mechanical. In this last category are Hall Effect switches (a magnet actuates a semiconductor); capacitor-coupled switches; conductive elastomer switches operated from the membrane keyboard; and others.

Keyboards may be replaced one day by voice communication, as the techniques for producing sounds are perfected and costs are reduced.

Printers

Permanent records of computed and displayed information are created by printers in response to binary strings of information from the digital control device processors. A printer must be matched logically to the device from which it receives information. It may be designed or programmed for serial or parallel transmission of data, and rates of transmission must be established to satisfy the slower mechanical action of the printer with the almost instantaneous data transmission rate.

The method of forming the printed character is a function of the purpose of the printed material. The character created by a printing mechanism may be solid, the result of impact from an embossed ball, metal tape, or lever-driven rod. More commonly, because the printing head is less expensive, it may be formed from a matrix of dots, with the printing accomplished by impact of rods through an inked ribbon, or by non-contacting rods. These may be heated to activate heat-sensitive paper, or they may create a capacitive or electrostatic charge, again for activation of special paper surfaces. The character type produced by impact is usually used for the best quality of printout. This is often referred to as "letter quality," the type that would be used for business correspondence. Laser actuated printers have been developed and are used for creation of multi-copy printing where speed is important. This is more common for office and public relations applications than for digital control and industrial records production.

External Memory

Distributed control systems and minicomputers use floppy disks and hard disks to supplement RAM memory storage. Mainframe computers use tape on reels for the same purpose. This extends the solid state memory in the digital control device, which is limited, and at the same time protects information by providing a backup storage capability external to the digital equipment.

Disks and tape have a magnetic surface, used as a storage medium. The surface moves in relation to a very small electromagnet that generates pulses to create magnetized areas. These can be read as binary bits from pulses produced by another electromagnet as it passes a magnetized area and transmits the stored information to the digital equipment memory.

Two types of disk drives are commonly used: the floppy disk and the hard disk. Figure 4-5 shows the mechanical arrangement of a floppy disk. A flexible plastic sheet coated with magnetic iron oxide is permanently housed in a square jacket with openings providing access for the magnetic heads. When placed in the disk drive, the disk spins at 360 rpm. The surface is formatted into concentric tracks, and each track is divided into sectors. Part of the first track is used as an index for defining the files stored on the disk. During a data transfer, the read/write head is positioned radially at the desired track, in near-contact with the disk surface through the access slot.

"Floppies" vary in size and storage capacity. Standard diameters are 5¼ and 8 inches. Storage capacity varies from less than 100K bytes for the smallest single-sided single-density disk to several megabytes for a double-sided, double-density 8-in. disk.

Floppy disks are inserted into the disk drive as they are used. The hard disk is a permanently encapsulated aluminum plate with a magnetic coating, and it is not removed. Hard disk drives, sometimes called "Winchester drives," can store many megabytes of information. 256 megabytes of storage is now available in 5¼ in. high, ½ rack width units. It is possible to stack platters in a hard disk assembly, up to four at the time of this writing.

Laser disks are still in the future, but it is expected that they will soon be introduced. They are said to be less expensive than the present form of hard disk, four times as fast, and will store 1600 megabytes. That is the equivalent of about 2000 average-sized books.

4-12 Power Backup

Like an automobile without a battery, a digital control device or system will not run without power. When mainframe computers were first used for process control they were not as reliable as they have since become, and the reputation for losing a system because the computer has "crashed" still persists. Reliability has improved, partly because solid state

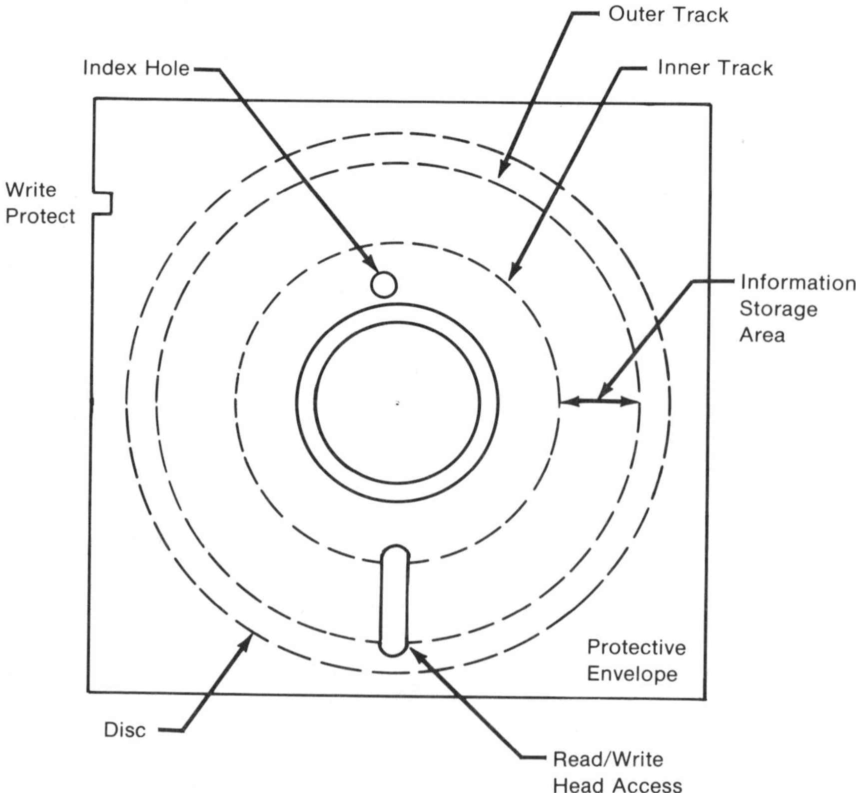

Figure 4-5. Floppy Disk

electronics generate less heat in operation, partly because quality control programs are enforced, partly because equipment is "burned in" (operated after manufacturing is completed and before shipment) for long enough periods to uncover a majority of incipient failures. Digital equipment should be designed to have redundancy of critical components. There can be software backup, whereby entire modules will be replaced by standby modules if a dead man timer times out before certain routine check points in repeatedly run background diagnostic routines are confirmed. Hardware devices can be used for backup, in the form of manual control stations and standby or concurrently operating analog controllers. Modules are designed so that most failures can be quickly corrected by replacing a single printed circuit card. One of the keys to keeping mean time between failures (MTBF) large and mean time to repair (MTR) small is to maintain an adequate supply of spare cards and to have personnel who know how to use them.

Nevertheless, in some processes down time can be so damaging to production that there must be no failure whatsoever. When this is the case, provision in the original system design **must** be made for some form of uninterruptible power supply (UPS). A UPS backup operates from the main ac supply to maintain a charge on a standby storage battery. If the utility power line voltage drops below a preset level, an inverter is switched into the battery output and maintains power to the electronic system until the battery charge is exhausted. Sufficient capacity must be provided for whatever period is critical to operation.

Approximately 10% of the total system equipment cost should be estimated for this type of backup. How important this is must be a management decision, made on the basis of production loss expense in the event of a failure. For a heat treating furnace control system, where heat will be retained in steel ingots for several shifts and no material loss can be incurred, backup may not be necessary. For a petrochemical installation where pipe lines will harden, full of thermosetting plastic, if flow stops for more than a few minutes a considerable expense can be justified for uninterruptible backup equipment.

EXERCISES

1. What considerations should be given to the design of the ac power source for a system using digital control devices?

2. What considerations should be considered when designing a central control room?

3. How are software and instrumentation documented on engineering drawings?

4. For designing and implementing a system, what should be the makeup of a project team?

5. What is meant by the terms common mode and normal mode?

UNIT 5
Communications

Digital control systems have become economical in part because of the cost reduction and the powerful applications that large scale integration and microprocessors have made possible and in part by using advances made in communication techniques that provide the means for moving large amounts of information long distances in a very short time.

In a system including field mounted equipment and digital control equipment, there are four categories of communication:

(1) Communication between field mounted analog equipment and the digital control portion of the system

(2) Communication between field mounted discrete digital equipment and the digital control portion of the system

(3) Communication between subcomponents within the digital control device

(4) Communication between digital control devices

The first and second forms have been discussed in Unit 4 and the third form in Unit 3. In summary, process variable values are measured. Analog signals, usually in the form of 4 to 20 milliamperes, or converted to voltages (1 to 5 volts dc, or 0 to 10 volts dc) are brought to the I/O terminations. Voltage wiring will normally be twisted shielded pairs, with the shield connected to ground at the I/O termination end only. Sixteen gauge wire is a good compromise between size and strength. DC signal wiring must be physically separated from ac wiring. If there is radio frequency interference, wires may have to be run in grounded metal conduit. Care must be taken to eliminate ground loops and induced voltages from capacitance coupling or transients, and to have one and only one system common point. This should be grounded separately from the power ground.

Discrete digital values, usually ac or dc voltages, can be brought to I/O terminations over single wires of relatively heavy gauge (14 or 16 AWG).

Digital devices communicate between themselves by sending energy representing 1's and 0's over communication links. Moving 1's and 0's has a simplicity that is deceiving; there are many aspects to consider, and many terms used may be unfamiliar to the reader. Indeed, there are so many such terms used in writing about communication that it will be well to begin with a glossary of those used frequently in the text, as a frame of reference.

5-1 Glossary

asynchronous transmission. Serial transmission of the bits making up a single character, preceded by a "start" bit and terminated by one or more "stop" bits.

bandwidth. The range of frequencies a circuit will pass. The faster the digital signal transmission rate, the greater the bandwidth requirement.

baseband. Transmission of signals without modulation. In a local network normally used for digital control devices where 1's and 0's are inserted onto the network as voltage pulses, the entire bandwidth of the cable is used by the signal.

baud. The number of changes per second of signal level taking place on a transmission line.

bisynchronous transmission. Byte serial, half duplex synchronous transmission.

broadband. Transmission media capable of passing wide bandwidth signals. On local networks, broadband refers to the use of frequency multiplexed cable. Broadband transmission is not normally used for digital control device applications.

bus. A physical path for passing signals from one point to another.

carrier. A radio frequency or light wave that is transmitted over a cable and modulated to produce a transmission.

crate. A term used in the CAMAC* standard. A remote I/O housing containing modular I/O equipment including A/D converters, microprocessors, etc.

crosstalk. Contamination of a signal in one channel, by spillover from another channel of information.

data highway. A single conductor transferring information between nodes.

DCE. Data communication equipment. The sender in a RS-232C** transmission.

DTE. Data terminal equipment. The termination point in a RS-232C transmission. A terminal can send or receive data.

format. The arrangement of binary information making up a frame.

frame. A block of information , preceded by a header and followed by a trailer to include address and control information. Frequently used in serial transmission.

full duplex transmission. Transmission that is simultaneous in both directions.

global highway. A combination of local highways and data highways making up a complete communications system including a number of Local Area Networks (LAN).

half duplex transmission. Devices take turns transmitting; transmission may be from A to B, or from B to A, but not simultaneously.

handshake. Communication between a sending device and a receiving device to establish that a message is ready to be sent, and acknowledging that the receiver is ready to receive it.

local area network (LAN). The network of communication links between the devices of a single system.

local highway. A single wire over which information is transmitted between devices grouped at one location. A local highway can interface to a data highway through a node.

node. A terminal location or switching point in a communication network.

packet. A string of characters that includes source address, destination address, a message, and control characters. In the process control context, a packet is similar to a frame.

parallel transmission. Movement of information, four, eight, or more bits simultaneously.

protocol. A formal definition of data formatting, error checking, priorities, and device assignments.

serial port. An input/output gateway that transmits data one bit at a time.

serial transmission. Movement of information one bit at a time.

simplex transmission. One device transmits; transmission is in one direction only, from device A to device B.

synchronous transmission. Data being transferred is in synchronization with a timing signal.

5-2 Information Transfer between Digital Control Devices

Within the electronic device itself, over very short distances, bits of information will usually be moved in parallel. Between components on printed circuit cards this will take place over buses — parallel strips of foil. The cards may plug into a backplane, which

---

*CAMAC. Acronym for Computer Automated Measurement and Control. The name of a modular instrumentation standard for interfacing transducers and other devices to digital controller.

**RS-232C. A standard created by the Electronics Industries Association defining electrical characteristics for an interface between terminals for time/share use and a modulator/demodulator for encoding of digital data into voice-like signals on telephone systems.

interconnects circuits using buses of foil. Between devices, when the distance is only a foot or two, flat ribbon connectors may be used for parallel transfer of information. Figure 5-1 shows the bus-backplane-ribbon cable information path.

Parallel transmission is very fast, but it is practical only for short distances. It will often be used to pass large amounts of information between devices that are only a few feet apart, such as a computer and a printer, or a disk drive.

Keeping track of all the bits in a byte moved in parallel over a long distance and putting them back together into a message (particularly if there had been any branching or repeating along the way) is difficult and expensive. More practically, a single conductor costs less than eight or sixteen conductors. Therefore, the method of transmission when distances are longer than a few feet is to send bits serially, over a single conductor. Conversion from parallel to serial transmission, described in Section 3-8, is illustrated in Figure 5-2.

The following transitions are shown:

(1) Bits forming the byte 01101001 are moved serially into a register full of 0's, pushing them out. The 0's may be lost or overflow into another register.

(2) The bit formation 01101001 moves in parallel to another register, leaving the same configuration of bits behind in the register where it had been stored.

(3) 01101001 is moved serially out of the second register onto a bus, perhaps for transmission through a port. The register remains loaded with the 0's used to displace the formation of bits.

Transmission can proceed in two directions. Full duplex transmission is normal in telephone communication and may be used on interoffice networks. Digital control systems may use full duplex but will usually operate in a half duplex mode.

5-3 Local Area Networks

A local area network (LAN) interconnects a system of digital control devices in an operating area and functions as a conduit to convey information between devices. A network is made up of hardware at the network nodes where devices are connected; the cabling between nodes, over which the information is transmitted; and the communication protocol

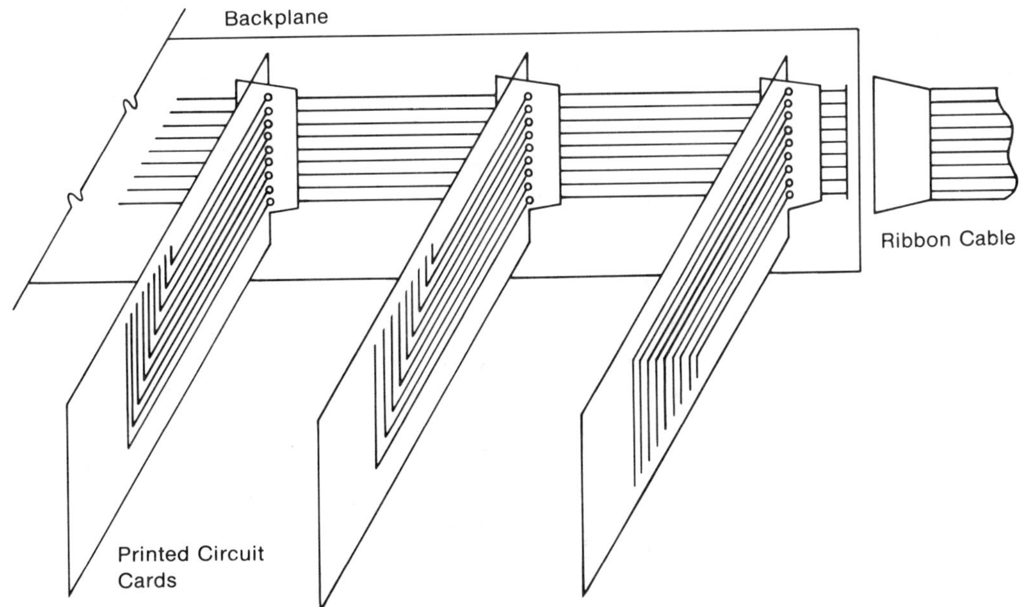

Figure 5-1. Parallel Transmission over Buses

that controls and directs the data transmission. Local area networks, in turn, can be combined with other communication nets in a global network to transfer information over an entire plant or plant subdivision area. Various departments of a plant can share manufacturing information electronically or a LAN. Having current data available to all users on a real-time basis implements improvements in management decision making, promotes energy management programs, and results in significant cost reductions, productivity improvements, and schedule simplification. This concept is considered in the discussion of hierarchies in Unit 13.

Figure 5-3 illustrates the common forms of local area networks. Networks may have a point-to-point configuration, connecting two components; or a star configuration, in which cables radiate from a central operation station to connect a number of remote operating devices. They may be linear, forming a data highway to which other satellite networks are connected. The linear highway may be bent into a circle, forming a ring network. Point-to-point networks and star networks are relatively short; information is carried between devices over shielded multiconductor (twisted pairs) cables 1000 feet long or less. Data highways, on the other hand, may be several miles long; the cable used is usually coaxial metallic conductor cable, or fiber optic cable. Highways may be hybrid; that is, the main highway may be made of fiber optic cable, while the satellite highways may be made of metallic conductor cable. Special devices are required to tee into a fiber optic cable from a metal conductor cable.

The star configuration has the limitation that the central operating device is a common point in the entire information transfer, and its failure can shut down the entire network. In addition, as the number of connections increases, the memory of this device must be used for more and more message switching, and there will be less remaining for computation and control. A linear or ring-type data highway will run through nodes to which various devices are connected; or other linear highways may feed into a data highway through connecting nodes. Transferring the message switching activity to nodes on a linear or ring highway improves the network efficiency. Failure of a node will shut down only part of the network; it is common to use redundant highways and nodes, greatly improving reliability. The computation and control activity can be concentrated in a single device, located centrally and communicating over the network with satellite devices.

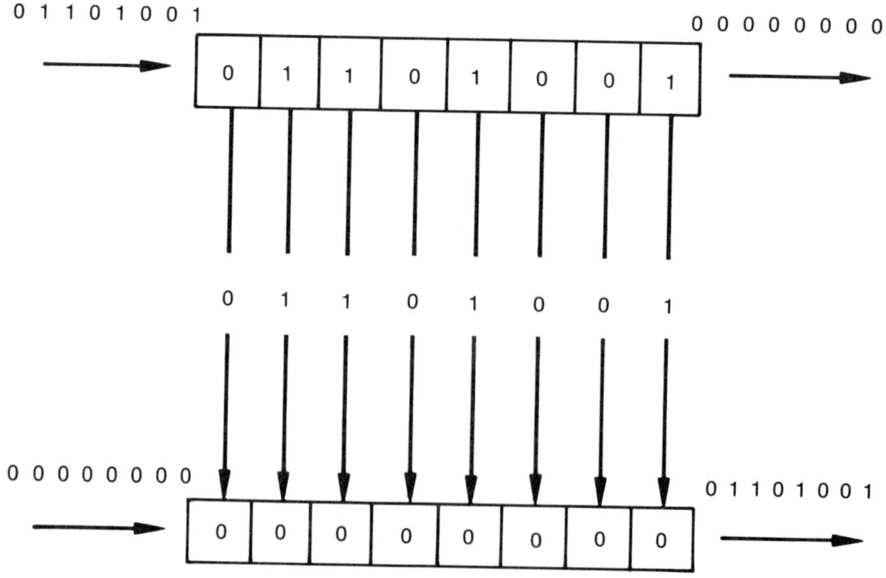

Figure 5-2. Data Transmission and Conversion: Serial-to-Parallel-to-Serial

5-4 Nodes and Connections

The shielded multiconductor cables used for point-to-point and star networks have only two ends and terminate in multiconnector plugs and sockets. Data highways that have a number of branches will require hardware so that branches can tee into the main highway. This is not difficult to do for metallic connectors; tees and branches for coaxial cable are available at relatively low cost.

Teeing into a fiber optic cable can be done in a number of ways, none as simple as making a solder connection or using terminals. Optical multiplexers use software to select one of a number of fibers that are all optically connected to a common cable; light can be

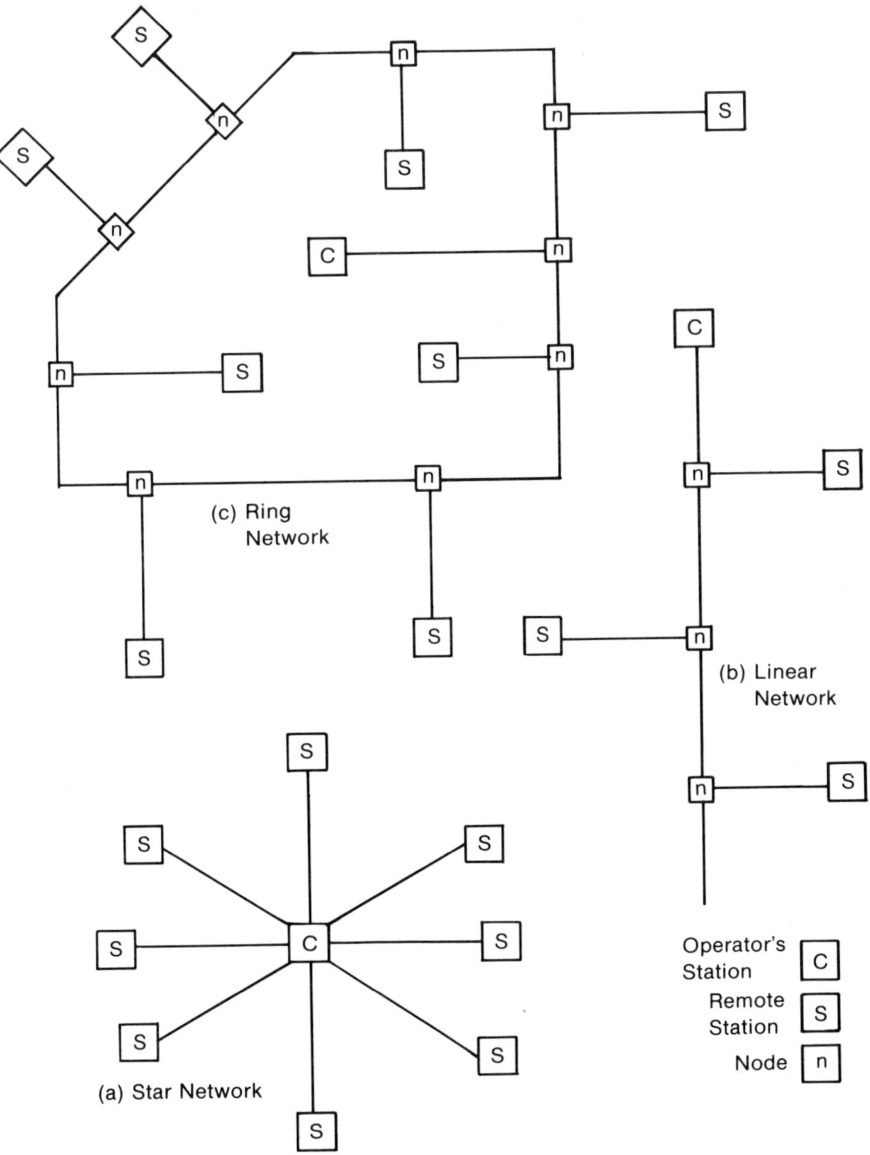

Figure 5-3. Local Area Networks

changed to a low level voltage using p-i-n* photodiodes, amplified, and changed back into light energy again through an LED and a photodiode circuit. A converter of this type is shown in Figure 5-4. The arrangement is typical for repeater interfacing of fiber optic cable or for interfacing fiber optic cable with metal conductor cabling at active system stations.

Signals do not maintain their strength or purity indefinitely, and amplifiers for repeating and strengthening signals must be part of a long network. Amplifier connections for coaxial cable can be made using standard connectors; for fiber optic cable, the optical signal must be changed to electrical energy, amplified, and used to light an LED, as described above for tee connections.

5-5 Data Highway Material

Communication links that interconnect devices in a network should be reliable, easy to install and easy to maintain. Copper and aluminum metal conductors are used for the majority of installations. Where problems are created in maintaining transmission in metallic materials because of electrical interferences or corrosive environments, fiber optic cable is finding use because it can function under these adverse conditions. Fiber optic cables are also advantageous when the amount of data that must be transmitted becomes very large, because fiber optics has large information carrying capabilities.

Twisted pairs of metal wire in gauges from 18 through 22 are easy to install, connect to, and add to. The numbers refer to American Wire Gauge standards; the lower numbers are the larger in diameter. The recommendation was made previously to use 16-gauge wire in order to have enough pull strength. That recommendation was for pairs of wire, used for transmitting signals from field-mounted equipment. Many pairs of the smaller sized wire

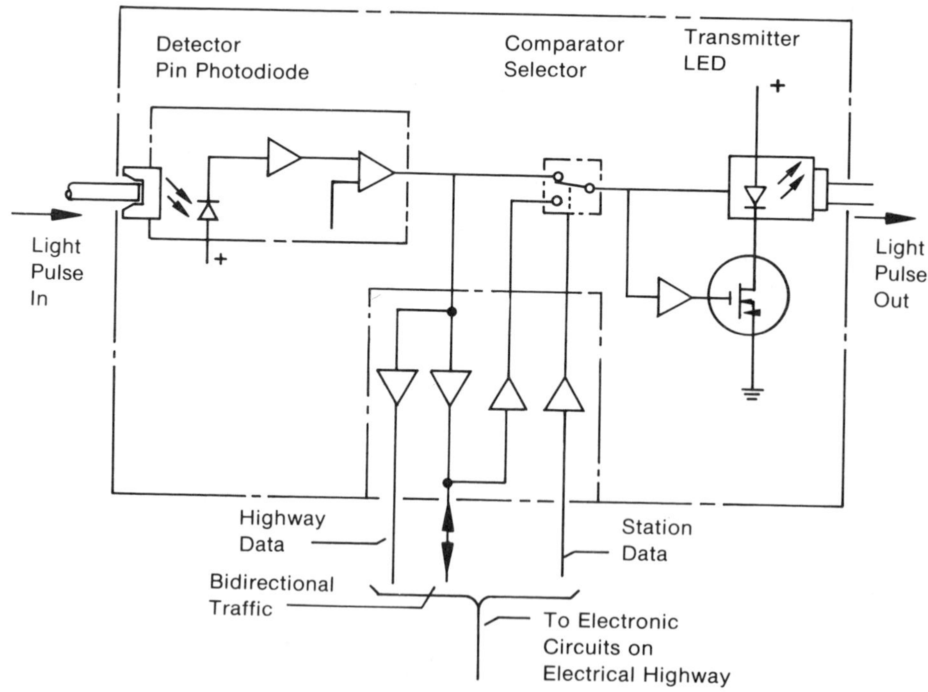

Figure 5-4. Fiber Optic-to-Metal Highway Connector

---

*p-i-n. A p-i-n diode has an intermediate intrinsic section, a depleted region with a high electric field that separates the n and p regions.

may be combined in a single cable for interconnecting the digital devices. The cables used in star configurations, for example, may be multi-conductor, made up of shielded pairs of 22-gauge wire.

Costs of installation of twisted pair conductors are low if wiring troughs can be used ($.30 per foot is typical of cost in some parts of the U. S.), but if local codes require wiring to be run in conduit, costs become high. Capacitive coupling between adjacent wires attenuates the signal in this kind of wire and signals lose power. This limits the length of effective run to about one km, but the amount of traffic that can be handled effectively over this distance is as much as one million bits per second.

Coaxial cable, commonly used for the data highway, is usually made up of one heavy central connector separated by an insulating layer of plastic or foam from an outer sheath of braided copper or aluminum. Installed costs will be between one and two dollars a foot, depending on the size of cable and local codes. It is not easy to install because it is fairly stiff and cannot take sharp bends. Care must be taken to not nick the cable or put kinks in it.

On the plus side, the shield helps to eliminate much of the electrical interference that bothers twisted pair installations. It is particularly effective against electromagnetic interference from sources such as radio transmitters. It can carry a lot of traffic, probably as much as ten times that of an equivalent sized twisted pair.

Figure 5-5 shows the construction of coaxial cable, twisted shielded pair cable, and fiber optic cable. In (a) the braided shield of the coaxial cable provides excellent electrical field protection but only fair magnetic shielding. The foil shield of the shielded twisted pair cable in (b) provides better coverage and more effective magnetic shielding than a braided shield but has less flexibility and durability. The optical portion of the fiber optic cable in (c) is wrapped with tough Kevlar™ to provide pull strength, which can be as much as 70 pounds for a Kevlar-wrapped fiber 100 microns in diameter.

Fiber optic cable is not affected by electrical interference as are coaxial cable and twisted pair cable; the signal passing through the fiber optic cable is made up of pulses of infrared and visible light. The cable can be made with such purity that tremendous amounts of data can be transmitted for long distances before the inevitable dispersion from small impurities requires that the signals be regenerated by amplification. A typical cable used for distributed control system highways has a diameter of 100 microns and carries one million bits per second for a distance of 7000 feet before the signal has to be repeated.

This kind of cable is made of an inner silica glass core surrounded by a glass or acrylic cladding. The difference in refractive index between the glass on the inside and the cladding on the outside assures that light entering the core at a shallow angle will bounce back from the cladding and be sent down the inner tube as if through a guided pathway.

The installed costs of fiber optic cable are about the same as for coaxial cable. When branches have to be made, or satellite wire conductor highways have to be connected to a fiber optic highway, however, the cost of the fiber optic installation increases rapidly. Fittings exist for connecting one length of fiber optic cable to another, but it is not practical to tee directly into a cable.

5-6 Information Transfer — Protocol and Format

Communication is much more involved than simply connecting devices together and sending messages. A number of levels of information transfer must be considered.

First of all, the information to be transmitted has to be selected and defined. Then it must be moved from its memory location to a port, converted into a form that will let it be sent on its way, and formatted in such a manner that it will reach the right destination and the receiver will be able to interpret the message. It must carry with it the means to verify that it was received correctly. Finally, it must be stored in a specific location so that it can be used for whatever purpose dictated the request for its transfer in the first place. All of this is done by combinations of hardware and software.

Information in the computer is stored in registers and usually moves out of them in a parallel configuration. When the system recognizes that information is wanted, it interrupts its normal business to get it. In order to get the digital information from the registers onto the link in serial form, it must be sent through circuitry that changes the configuration from parallel to serial and formats a message so that it will be directed to a particular destination and received there. Typical of solid state devices that do this is the UART (Universal Asynchronous Receiver/Transmitter), in combination with a modem. At the transmitting end of a link, the UART receives the signal in a register and moves it out one bit at a time to create a serial string. The string of bits is combined with other bits to form a header that announces that a message is coming. At the end, bits are added to announce that the message has ended. At the receiving end of the link another UART strips off the code bits and packs the serial message into a receiving register. From here it passes over a bus in a parallel configuration to its destination.

A UART will often be built into a single chip. Figure 5-6 illustrates a standard UART, consisting of a transmitter, a receiver, and control logic. An external clock is synchronized with the serial signal to establish the bit period. The control inputs are parity inhibit, odd/even parity, stop bit select, and word length select.

UARTs were developed for asynchronous transmission, although there is a USART that supports bisynchronous systems. In asynchronous transmission the information is divided into characters of fixed length. An example is a single ASCII character preceded by a START code and followed by an END of message code. When data is not being transmitted,

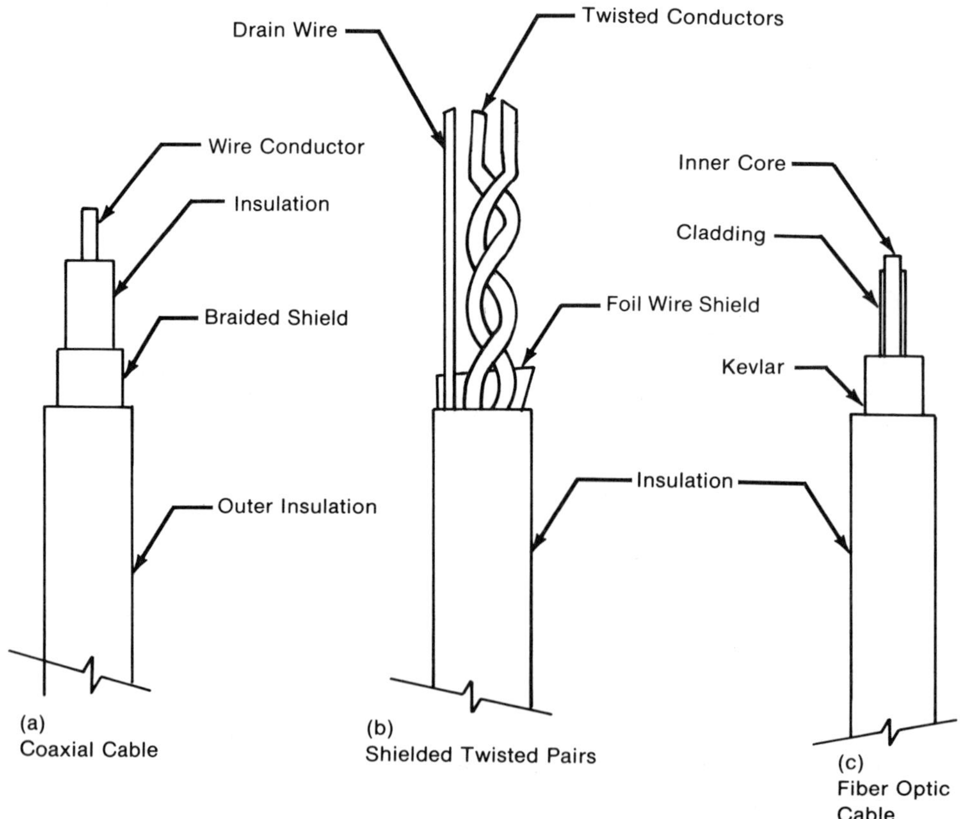

Figure 5-5. Materials Used for Data Highway Cables

the line is held in the binary 1 state. This will be some specific voltage level, although that level may be zero volts. When a character is to be transmitted, a binary 0 is sent for one bit time before sending the other bits. This is the START condition, and when the receiver detects the leading edge of the START bit it begins counting time periods. It knows that an END of message bit should be received after a certain number of counts.

Data communication uses the terms "mark" for sending a binary 1, and "space" for sending a binary 0. A transmission line is said to be marking between characters in asynchronous transmission. The format of a character sent by asynchronous transmission is shown in Figure 5-7.

Between START and END the transmitter watches a clock to send signals at specific time intervals. Baud rate is established by the clock inputs to the transmitter and receiver. If ten 11-bit characters are transmitted per second, the baud rate will be 110. Standard baud rates are 50, 75, 110, 134.5, 150, 300, 600, 1200, 1800, 2400, 3600, 4800, and 9600. Telephone transmission is usually at a rate between 110 and 1200 baud. A common rate for information passing from personal computers to printers and through modems is 300 baud.

The receiver uses the same time interval to expect receipt of a bit. The time interval establishes the relationship between bit rate and baud rate. Suppose that 5 volts corresponds to a 1 bit. At 300 baud the voltage associated with a single bit exists for 3.33 milliseconds. If the same voltage lasts for 6.66 seconds at this baud rate, the receiver knows that it has received the bits for 11.

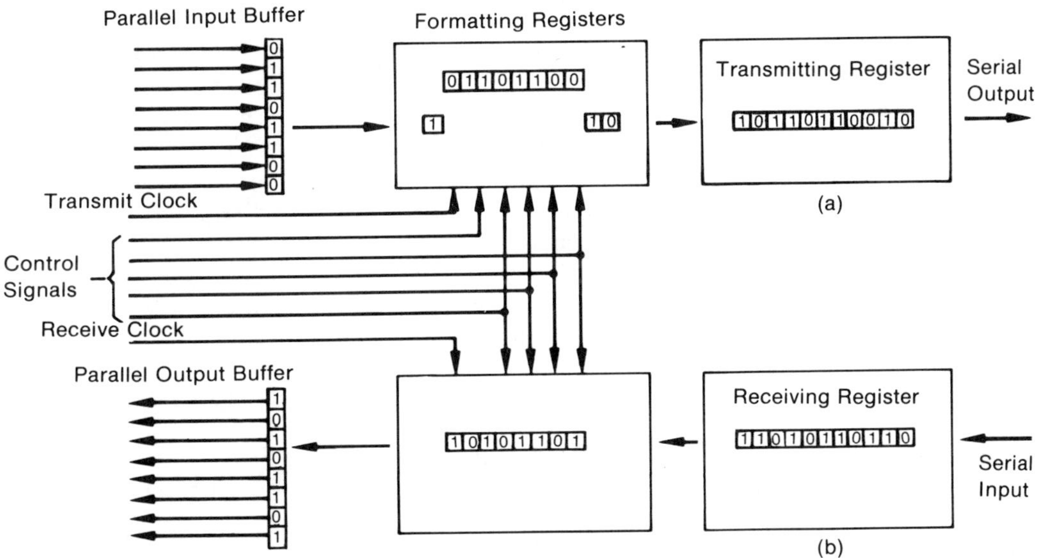

Figure 5-6. Standard UART: (a) Transmitting, (b) Receiving

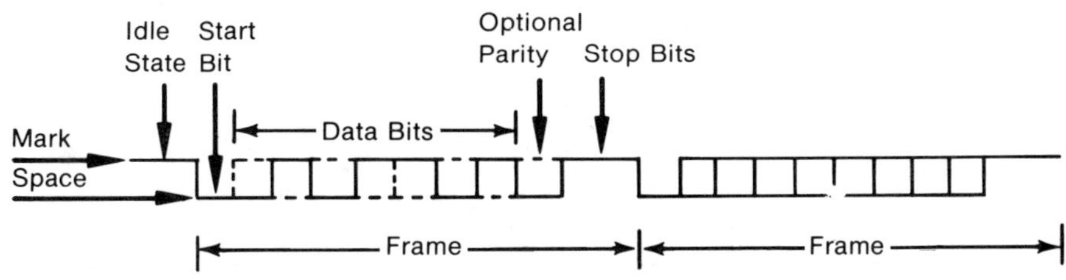

Figure 5-7. Asynchronous Format

The statement commonly made that baud rate is approximately the same as bit rate cannot be interpreted to mean that it is the same as information rate. Some of the changes in state during a single character transmission are START and STOP coding, and not message bits. Other portions of the transmission of a character may include READY TO SEND and READY TO RECEIVE messages, which make up a "handshake" that clears the way for an authentic transmission. Also, in a half duplex system there is turnaround time for a receiver to become a transmitter, and vice versa.

Asynchronous transmission is not very efficient, since about 40% of every transmitted character may be coding while the rest is actual message. Asynchronous transmission was fine for teletypes, which sent one letter at a time and had to wait for the keyboard to be struck before a new character appeared at the UART. Baud rates were low; 300 baud was commonly used, and 9600 baud was a fast system. Data highways must have much greater baud rates in order to move information from many sources to a central operating station. Some distributed control systems advertise baud rates of 2 million over their data highways.

For this type of information transfer, synchronous transmission (much faster than the asynchronous variety) is used. To increase the efficiency of transmission, messages are formatted differently, using a bit arrangment that sends a large block of data, preceded by a header and followed by a trailer. Eighty to 90% of the transmission consists of real data. The data being transmitted is in step with a timing signal or clock pulse. The synchronizing signal notifies all receivers that data is ready for input, and each receiver "listens" for a code that identifies it as the destination of the message. A handshake is not required; the receiver locks in phase with the transmitter and receives one character after another until the trailer announces that the "frame" containing the message has been completely sent. The circuit performing the functions of formatting will be more sophisticated than the UART described above.

Synchronous transmission frames assume a number of different forms, depending on the supplier of the equipment. A protocol used by a great many suppliers of digital control devices used in networks is called High Level Data Link Control (HDLC) developed by the International Standards Organization (ISO). Figure 5-8 shows the frame format of a message using HDLC. This protocol does much more than transmit information. An idea of the power contained in the formulation of the information packet is indicated by a listing of the type of system instructions that can be included in the coding of the frame.

STATUS — request the status of logical states.

DEMAND — request values of numeric variables to be sent or received.

ARRAY — request arrays of variables to be sent on an event basis.

COMMAND — send logical or numeric data to another device.

DATA LOG — request numeric data to be collected at specified intervals, processed, and returned at other specified intervals.

Other formats used by digital control equipment suppliers include the following:

BISYNC (BInary SYNchronous Communications Message Protocol), developed by IBM.®

DDCMP (Digital Data Communications Message Protocol), used with the DEC PDP/11® series of minicomputers.

Flag	Source/Destination Address	Control	Information Packet Field	CRC Frame Check	Flag

Figure 5-8. HDLC Protocol Format

SDLC (Synchronous Data Link Control), also developed by IBM.

ADCCP (Advanced Data Communication Control Procedures), developed by ANSI (American National Standards Institute).

5-7 Information Transfer — Ports

The formatted message leaves and enters a digital device through a port — in reality a register that has an address, followed by hardware and software for interfacing to the data link. A number of specifications for ports set standards for external voltage levels and provide a description of the connections and how they are to be used. The most common standard specification, developed by the Electronic Industries Association, has the identifying number RS-232C. RS-232C relates to single-ended signals; that is, one side of the signal amplifier is at ground potential. Another name, meaning the same as single-ended, is unbalanced transmission. It is designed for short runs over wire cable with a single ground return. Because of the high impedance and unbalanced interface, it is susceptible to noise. Voltage levels may not be maintainable for more than 50 feet, and trans mission in accordance with this standard is usually limited to cases where distances between transmitter and receiver are 50 feet or less. Using large wire, lower baud rates, special shielding, and routing to avoid interference, longer distances up to 1000 feet can be achieved. However, other standards exist, and are recommended, for communication over distances greater than 50 feet.

The RS-232C standard performs a logical function, which defines the voltage levels of the states of logical zero and logical one. Logical zero is frequently higher in voltage than the logical one, but it may just as easily be represented by a negative voltage. The sending and receiving device circuitry will determine actual values. The standard also defines a physical function, assigning uses for the connector pins where the link plugs into the digital devices. Pins are assigned command functions, such as Request to Send, Cleared to Send, Ground, and Transmitted Data. Figure 5-9 shows the interchange equivalent circuit defined by the standard and lists the electrical parameters specified in the statement of the standard.

RS-232C addresses communication in terms of data communication equipment (DCE) and data terminal equipment (DTE). When the standard was first written, communications equipment referred to a transmitter (for example, a modem); terminal equipment referred to a receiver (for example, a teletype machine). Today, the same standard is being used where communication may be between two computers, with half duplex transmission. Each end of the line may switch from being receiver to being transmitter many times a second. For this

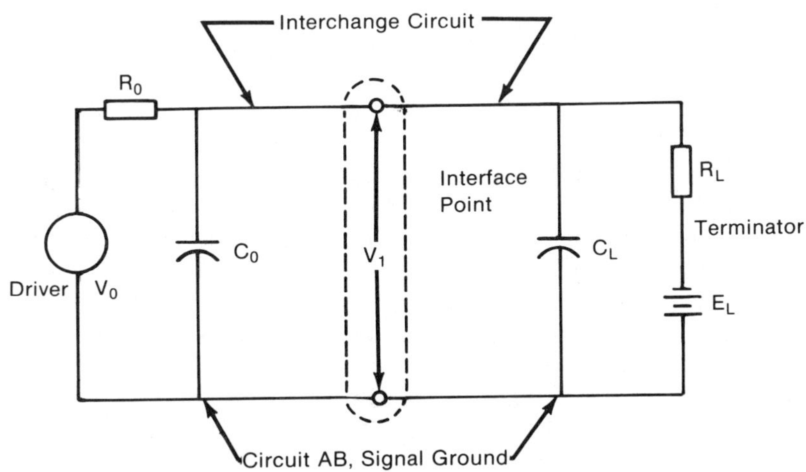

Figure 5-9. RS-232C Interchange Equivalent Circuit

reason, more is involved in specifying compatibility than simply having an RS-232C port at either end. In order for transmitted information to be received and understood, attention must be paid to the use of the conductors that carry signals to the ports.

For transmission distances longer than 50 feet, new standards have been developed and adopted. RS-449, for example, in conjunction with RS-422 and RS-423 is compatible with RS-232C and will eventually replace it. RS-499 is used for the physical portion of the standard, defining pin designations and connector requirements. RS-422 is the logical portion of the same specification, covering electrical signal characteristics that relate to unbalanced (receiver and transmitter amplifiers measure the differential between input and ground) line drivers. RS-423 is the logical portion of the specification, relating to balanced lines. In balanced transmission, receiver and transmitter amplifiers measure the differential between inputs. Interferences tend to cancel each other, contributing to the ability to transmit accurately over longer distances.

ASCII is often associated with RS standards because it was originally designed to be the protocol for teletype communication. The original function of the RS-232C standard was to standardize transmission between sending devices such as teletypewriters and receiving devices such as modems. Information is often stored in data bases and moved around through registers in ASCII format, so it is natural to move it from memory for transmission in that form. It is usually transmitted asynchronously. The same ports, however, with some hardware and software modification are used for synchronous transmission of packets formatted in HDLC, SDLC, etc.

IEEE 488 is another standard used to interface hardware devices such as digital voltmeters, oscilloscopes, spectrum analyzers, plotters, and so on. It was developed by the Hewlett-Packard Company® to ensure compatability between the company's many products and is used today by a large number of other manufacturers. Eight bits of data and control information are transmitted in parallel over a 16-wire parallel bus. There is detailed handshaking and a designation of receiver and transmitter functions. Although it is primarily an instrument interface, it can be used for process control.

Still another interface is the CAMAC standard, originally designed for nuclear instrumentation laboratories. The standard defines the interconnections between a computer and "crates" through a parallel bus. There is also a serial version. A crate is a relay rack-type framework for printed circuit cards with modules for A/D converters, microprocessor controllers, CRT drivers, and so on. Two cards in the crate handle the communication control. Crates can stand alone, be connected in groups of up to seven units to interface in parallel transmission with a central computer, or interface up to 62 at a time with a central computer over a serial highway.

5-8 Modems

Another link is needed in the communication chain because pulses of energy leaving the ports of digital control devices are at a low level and cannot travel very far. In order to survive they need a boost. This is provided by the modem, a device that accepts a binary coded message from the port and puts it into a form that will travel over the data highway. Modem stands for modulation-demodulation. There are modems that produce radio frequency signals, light signals, audio signals, or voltage signals. Voltage and light are commonly used for digital control system transmission. The modem modulates the transmission, sending a signal to represent a 1 or a 0 by changing the frequency, amplitude, or phase of energy level of a carrier. In frequency change, often used for digital control systems, a square wave changing from 1 volt to 2 volts ten times a second might represent a 1, while the same change 20 times a second might represent a 0 condition. This change of energy level would be converted into pulses of light ten times a second or twenty times a second if the transmission medium were fiber optic rather than coaxial cable. The modem at the receiving end reverses the process, converting the modulated signal back into a string of binary bits.

A modem designed for phase shift keying is often used when information is sent at baud rates faster than 300. This type of modulation reads the digital data and shifts the phase of its analog output when there is a bit transition from 1 to 0 or from 0 to 1. This is illustrated in Figure 5-10. When the receiving modem senses a phase shift in the analog signal, it reverses the conversion and generates a 1 or a 0, depending on the direction of the shift.

5-9 Driver Programs

In order to pass digital information from one device to another, both devices must speak the same language; having formats and protocols assures that a message will arrive where it should, but it does not guarantee that it will be understood. Putting a letter into a stamped envelope and dropping it through a slot at the Post Office presumes that it will be sent to the correct address; but it must be in a language that the recipient understands for the transmission to become effective. In the same way, for digital control system information to accomplish its purpose there must be a program written to set up the rules for the translation of information into a form that an application program can use. This procedure is called a driver program.

A driver program supervises all the transactions that take place between a device and the programs that call it. It makes the transfer of data transparent, so that an external device appears to the operating system to be just another port. It will translate between the language of the computer operating system and that of the device. It may include handshaking routines to announce the presence of information to be transferred, or the need for it; checking routines to test for accurate transfer; and routines to transfer blocks of information. It is the program that contains the codes for asking for information from the appropriate part of the data base of the computer, and for storing it into buffers that can be accessed by the external operating system. It is brought into operation by a command (usually CALL or INPUT or OUTPUT) from the application program written for the device controlling the transmission.

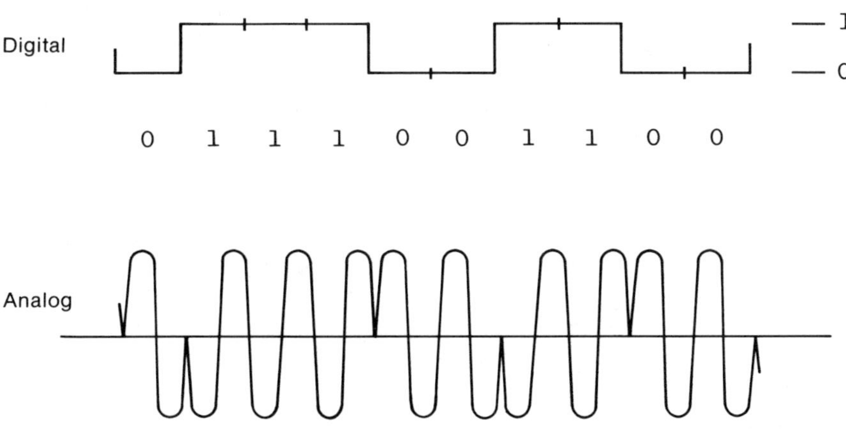

Figure 5-10. Modem — Phase Shift Keying

5-10 Information Transfer — Traffic Control

The primary purpose of life for any network is to move information. When there are many users, the managing of message traffic becomes more complex than for the most crowded automobile highways, although the rules for keeping traffic moving in an orderly and efficient manner are similar. There must be provisions for waiting in line, for finding a destination and letting the folks at home know that arrival was achieved safely, for maintaining the right of way, and for preventing slow components from blocking faster ones.

Communication time must be divided so that all devices can communicate in a very short time frame, preferably not more than one half second. There may be several thousand devices, and time must be divided into very small parts. This is not difficult; it is possible to measure accurately one millionth or even one billionth part of a second.

One way to use the time is to poll the devices, constantly monitoring to see who wants to communicate, then giving that device access for a specific number of milliseconds. Another arrangement has a schedule, where each device has a reserved slot of time. If the slot is not used, extra time becomes avilable for the busier of the remaining devices. Both of these methods require a master controlling device — a traffic director. This can be a point of vulnerability in case of hardware failure. Polling and traffic directors are often used with star networks and with linear highways.

Another method allows all devices to use the highway at any time. Part of its name, therefore, is "multiple access". Courtesy must be observed. If a station is transmitting, all other stations stop and listen. Stations can tell if another station is transmitting by listening for the system carrier frequency; this leads to the name "carrier sense". If two stations start simultaneously both must stop and the one with the lower priority waits for the other one to speak. From this, the final contribution to the name is "collision detection". The complete name, then, is Carrier Sense Multiple Access/Collision Detection (CSMA/CD).

Two methods used for ring networks involve passing a signal around the ring. In one method a message called a frame is passed from station to station. It can be likened to the use of envelopes in our postal system. The frame has leading and trailing information with a blank space in between for a message. If a station has information to send, it fills in the blank space as a frame goes by. All stations read the frames as they go by, but they remove and store the message only if it is addressed to them. When the empty frame arrives at its point of origin, the sender knows that his message was received.

The other method is called token passing. A token is a very short message that circulates, announcing that the highway is clear. Stations on the highway can use the token according to a configured sequence for a short period of time, typically 50 milliseconds. The station next in sequence, if it has a message to send, grabs the token and changes it to read "here comes a message". It attaches its message, which may be as long as the time frame allows, onto the token. It is removed at its destination, and the token message changes back to the "clear" format.

Still another form of network communication creates one all-encompassing data base for the entire system, sometimes called a global data base. The protocols discussed above operate like a private telephone system, with conversations from address to address. This type of network is more like a collection of bulletin boards, or the personal column in a newspaper.

Each element in the system includes a block of shared memory, which is available to every other element. Information generated as the result of processing input signals or of performing control operations as functions of deviation from set points is placed in the shared memory. Periodically, the information in the shared memory is broadcast over a data highway and becomes accessible to every processor in the system. If an element needs information from another element, it listens until it hears that element broadcast. Then it pulls the items of desired information off the highway and makes it a part of its shared memory. Each processor does nothing to communicate with the rest of the system and treats information as local to itself. In effect, each processor sorts out, from all the available information, that which is pertinent to it, as would a person reading a bulletin board.

The amount of highway traffic in a global data base system is large, and communication must be very fast. The Westinghouse Distributed Family Processing System™, a distributed control system that has used this form of communication, lists a highway rate of 2 megabaud. Forms of the method are used by the Honeywell TDC-3000™ distributed control system and by the Measurex Vision 2002™ system. Both of these systems are proposed for total plant control applications. This method of transmission has been advocated (Ref. 11) for hierarchical systems (discussed in Unit 13) as better adapted to an all-factory communication system than is a global network that is parochial, either to specific local areas or to specific categories of information.

Networks that use the time division techniques are called baseband systems. Like a single-lane highway, only one message is passing at a time. Broadband highways used for voice communication make use of a number of frequencies and allow multiple message use. The total bandwidth of the cable can be divided between various parts of the network. If a network has only one frequency, the modem translating information from the internal format of a device to the highway format will be a fairly simple fixed frequency device, operating on a single channel. For broadband transmission, a more sophisticated multi-frequency modem must be used.

Once the message has gotten onto the highway, it must find its way to the correct destination. For best highway efficiency, the most information that can be handled should be sent in the minimum time allowable. Messages are usually sent in packets, and a packet is made up of a header defining the type of message and the destination, the body of the message, and some means of verifying the reliability of the message received. The message may consist of a single character or of many words.

Routines must be included with the transmission to detect any failure to communicate properly, so that errors are not transmitted. The reliability may be as simple as a parity check, the simplest form of error checking. Parity means that the arithmetic sum of the bits in a group must be either odd or even. Using ASCII code as an example (ASCII represents characters with seven binary digits), if it has been assumed that the sum shall be odd but the 1s and 0s add up to an even number, an eighth digit, a 1, is added. If the sum had added up to an odd value, an eighth digit, a 0, would have been added. The receiver checks to make sure that all groups it receives have odd parity.

For communication using LANs, error checking will almost certainly be a very complicated procedure called a Cyclic Redundancy Check, where each message is divided by a polynomial and the remainders are summed, producing a very unique number indeed. In any case, as the message is received the same computation that was performed at the sending end is also performed at the receiving end, and the two results must agree exactly.

5-11 Standards

Standards exist to define the scope of the various communications functions. The structure of the network and its purpose for existence will make some functions more important than others, and all may not be used in every network. Regrettably, there is a proliferation of standards, arising from the fact that every supplier has done what he had to to get his equipment onto the market, with little or no regard for the way anyone else was doing it. The result is that, even within one company, digital control products are often not able to communicate with other products.

Standards have been generated by many groups, and each standard is valid for a part of the communication picture. Groups that have developed recognized standards include the Electronics Industries Association (the RS series); the International Telephone and Telegraph Consultative Committee (X and V series of recommendations for levels of data interfaces); the American National Standards Institute (ANSI); the National Bureau of Standards; the International Standards Organization (ISO); the P802 Local Area Network Committee of the Institute of Electrical and Electronic Engineers (IEEE); the International Telegraph and Telephone Consultative Committee (CCITT); and the International Electrotechnical Commission (IEC) working group IEC/SC65C/WG6 through the Instrument Society of America Committee SP-72.

In attempting to organize the information about standards, reference is often made to "layers" of communication. There is a model for such a protocol, proposed by the ANSI X3 Committee, similar to a seven-layered OSI (open systems interconnection) model proposed by the ISO and supported by the National Bureau of Standards. The lowest layer defines standards for sending information from one instrument and receiving it by another. The RS standards relate to this level. The message formats used for passing information between stations are a function of the next layer. The layer above that is involved with protocols for switching or routing messages — traffic management.

A protocol frequently used for LANs interfacing digital control devices is X.25, proposed by CCITT. This protocol sets procedures for gaining access to a packet-switched network. It defines characteristics for the first three layers, and is almost identical to HDLC at Layer 2. At Layer 3, it provides a virtual circuit service between devices connected to the network. Altogether, there are seven recognized levels. The upper four, which are concerned with language translation, tracking and maintaining multi-terminal communication, managing and controlling structured data, and managing applications of data, are not usually addressed by suppliers of digital control devices or systems.

A recent series of standards gaining in acceptance is the Manufacturing Automation Protocol (MAP), originally developed by General Motors Corporation. MAP supports the interconnection of a number of programmable logic controllers from different vendors on the same network. It is described as a 7-layered broad band token bus-based communication standard and incorporates standards developed by the U.S. National Bureau of Standards, the International Standards Organization, IEEE, ISA, ANSI, and the Electronic Industry Association and drafts under development by the International Electrotechnical Commission (based on the ISA work).

In particular, it makes use of the IEEE 802 committee work in developing standards that will provide a network user with a guarantee that a system that meets these standards will be supported by a number of manufacturers.

The IEEE 802.3 standard is a standard for CSMA/CD, describing how it works, defining the physical medium (a baseband coaxial cable), defining the protocol for transmitting data, and defining the physical connections that are to take place between the communications network and the cable. The IEEE 802.5 is a similar standard for token passing on a ring-configured system, and the IEEE 802.4 standard is a standard for token passing on a bus-type communication system.

REFERENCES

1. Leibson, Steve, "What are RS-232C and IEEE-488?" *Instruments and Control Systems*, January, 1980, pp. 47–53.
2. Eckard, Mark, "Tackling the Interconnect Dilemma," *Instruments and Control Systems*, May, 1982.
3. Washburn, Jerry, "Communication Interface Primer — Part I," *Instruments and Control Systems*, March, 1978, pp. 43–47.
4. "Communication Interface Primer-Part II," *Instruments and Control Systems*, April, 1978, pp. 59–63.
5. Damsker, D.J., "Base Control System Specification on Communications Structure," *Power*, June, 1983, pp. 69–72.

6. Weissberger, Alan J., "Modems: The Key to Interfacing Digital Data to Analog Tele-comm Lines." *Electronic Design*, May 10, 1979, pp. 82–84.
7. Derfler, Jr., Frank J., "Selecting the Right Modem." *PC Magazine*, January, 1983, pp. 224–233.
8. Derfler, Jr., Frank J. and William Stallings, "Local Networks: A Guide for the Per-plexed." *PC Magazine*, August, 1983, pp. 237–268.
9. EIA Standard RS-232C, August, 1969. Engineering Department, Electronic Industries Association.
10. Higham, John; Burton Kendall; and Merrill Gerdts, "The Data Freeway Network: A Coaxial Multi-Drop Management Control Bus." *Control Engineering*, September, 1981, pp. 103–106.
11. Taylor, William A., "Why Automatic Factories Aren't." *Programmable Controllers*, November-December, 1984, pp. 19-31.

EXERCISES

1. Explain how buses and a backplane establish communication between a number of printed circuit cards.

2. Compare coaxial cable and fiber optic cable in the contexts of cost and physical characteristics.

3. Compare the postal transmission system to the transmission of information over a data highway.

4. Describe the flow of information from a thermocouple, through a data acquisition unit, to a printer.

5. Explain how transmission symbols are differentiated from words in an asynchronous transmission.

6. Modems use three methods to differentiate between 1's and 0's. What are they?

UNIT 6
Data Acquisition

Devices for data acquisition were among the first examples of digital equipment developed, because the need for improving existing methods of keeping production and testing records was very great.

A familiar activity in operating plants has been the collecting of data by the departmental clerk. He will show up at least once a shift with his pencil and clipboard, exchange pleasantries and scuttlebutt with the operators, walk up to the instrument panel, and copy current values from the recorders and counters onto his record sheet. The records are taken to the shift foreman who, when he has time, will compute ratios, efficiencies, and production figures to be sent in turn to the supervisor's office the next morning. The information is old when it is seen by the ultimate user, and its accuracy is a function of the ability and integrity of the individual reading the charts. There is virtually no way to classify and recall specific events with any accuracy, unless the occurrence was very recent. The chance of finding a trend that happened more than a few days ago depends on human ability to remember, without bias. The limitations of this form of data collection are obvious, but for many years it has been the only way to get the job done.

Dynamic testing is another activity requiring data acquisition and the large amounts of data that may be generated during a test create their own special collection problems. In many cases large numbers of measurements must be observed simultaneously, often too many for one person with a pencil and a pad of forms. The solution in some cases has been to take photographs of the banks of gages and manometers that cover the walls of test chambers.

6-1 Data Collection

Where is the data used, and why? Collected data is needed to understand what is going on in a process, or in a test; it keeps track of all the important variables. It is an important record of what went wrong when there has been an upset or a failure in operations or tests. Collection of data for record purposes is vitally important if government agencies are concerned with the product. Everything that happens during the manufacture of a batch of food or pharmaceutical products must be available for review by the Food and Drug Administration. The Environmental Protection Agency requires comprehensive records of waste water and gas analyses so that it can monitor pollution. Other uses for collected data are improvement of quality, assurance of safety, and reduction of cost. Unquestionably, the acquisition of data is important to the performance of any plant or production department.

In most cases "data" refers to process variable signals, real-time analog values. On-off status of equipment can be equally important as data, and most data acquisition units accept a number of discrete inputs.

Acquisition and processing of large amounts of data can be done by minicomputers, but that use for a powerful machine is far below its potential. To use a computer for that purpose, a lot of programmimg time and money would have to be spent to develop routines for scanning inputs, multiplexing, conditioning signals, storing, processing data, and many other laborious tasks. This sort of activity reduces the computer's capacity to analyze and interpret. The demand for collections of data is sufficiently large that solid state electronic devices dedicated to data acquisition are commercially available, designed specifically to handle the collection work and signal preparation. Their circuitry will, at the very least, include signal conditioning with amplification, biasing, linearizing, filtering, and scaling of input signals to a common basis; multiplexers for selecting inputs; sample and hold circuits for maintaining consistency of readings during conversion; an analog-to-digital conversion circuit; control logic to supervise the scanning and multiplexing and conversion; memory, probably with battery backup, for storage of data; and ports for making use of the data for display, printout, or transmission to other devices. There may be a digital-to-analog conversion for generating output data.

Another approach is to use personal computers; special interfacing software and support software are becoming available to convert low cost personal computers into powerful data acquisition systems.

6-2 Data Logging

Thanks to microprocessors, the distinction between the activities of data logging, data acquisition, data processing and data reduction have become blurred. In this discussion, data loggers measure specific inputs on a time basis and print them out to a tape or strip chart recorder for analysis or record. Data acquisition devices measure data and compute on-line values that can be used for analysis and for real-time control. Data reduction means that the data is presented in an organized format, grouped chronologically and organized by subject. Data processing involves operations on the data that analyze it, compute new values, and change it to a form that is more meaningful than the raw data provides.

The problem with the definitions is that data loggers can do many of the things that data acquisition devices can do. Both types of equipment condition input data and can scale, manipulate, and check limits. To make a very broad generalization, data acquisition systems are faster and have broader capabilities than data loggers; they can be used for on-line data conditioning and reduction, permitting real-time control. The logger, on the other hand, is used primarily to collect and store information.

6-3 Data Acquisition Instrumentation

The simplest form of data acquisition system will be a self-contained unit to which all inputs are wired, and from which data is made available for viewing on an LED display or on a tape printout. It can be much more extensive, however. The other extreme is a system in which the acquisition and conditioning are done by modules packaged in housing suitable for harsh environments, even to the point of daily hosing down, as is required in some food processing facilities. These units can be located in processing areas where the data signals are generated. The data processing of an extensive system may be located in a central operating area, with peripheral printers, displays, and recording devices making processed data available to the user. Information will be transferred between the various discrete locations over communication networks, which may be data highways gathering information from modules located more than a mile from the centrally located processor. The communications may include modems and telephone transmission to provide connections to stations many miles away. If the data is used by a peripheral computer, a communication link will have to be provided for the transfer of information to and from it. Figure 6-1 illustrates typical data acquisition instrumentation.

Commercially available units will have some or all of the following capabilities.

- They can do the conditioning, linearizing, and converting for the following types of inputs:

    Thermocouples
    RTD's
    Millivoltage and voltage
    High voltage — 125 V dc, for example
    Pulse counting
    Contact inputs
    BCD

- They can run background programs for self-diagnostics, perform error detection and cold junction compensation. Some remote units add data processing capabilities and produce digital outputs.
- Packaging will be designed for normal as well as for harsh environments. Housings should conform to at least NEMA 12, and preferably NEMA 4 standards.
- 1000 or more input/output channels will be available, although modules may have to be combined to accommodate this much information.
- Typical operating specifications include the following:

  > Operating temperature — 0 to 50 degrees Centigrade
  > Humidity — 0 to 90%, noncondensing
  > Power — 115 V ac or 220 V ac (50 or 60 Hz)
  > Power consumption — 30 watts with printing, 190 watts
  > Resolution — 14 bits (1 part in 16,384)
  > Normal mode rejection — 70 dB at 60 Hz $\pm 0.01\%$
  > Common mode rejection — 140 dB at 60 Hz $\pm 0.01\%$, 120 dB at dc

Note: These are listed specifications for the Acurex AutoCalc™ data acquisition unit.

6-4 Data Processing

The centrally located unit will usually be the processing point for data, whether it is input directly or transmitted from a remotely located data logger. The information processor will be able do some or all of the following tasks.

- Conversion of data to engineering units
- Statistical analysis — averages, means, and standard deviation
- Display of selected critical points
- Generation of up to four alarms for each input
- Storing data
- Formatting and outputting reports

Figure 6-1. Data Acquition Logging, Processing, and Recording Equipment.

(Courtesy Acurex Corporation)

- Advanced mathematics — scaling, non-standard linearizing, Boolean logic, group averages, time averages, moving window averages, maximum values, minimum values, median values
- Steam table calculations
- Accumulation of data before, during, and after specific upsets or events, such as an alarm
- Outputting of data to computers and peripherals
- Communication, through modems, with remote site units (such as pumping stations), auto dial telephoning, transferring instructions to take data, and receiving it back at the central location
- Specific computations, using built-in algorithms (subroutines) for relative humidity, combustion efficiency, orifice flow, mass flow, tank volume, steam enthalpy, pressure and temperature compensation

6-5 Interfacing to Computers

A user who wants to access acquired or processed data for his computer-based application programs must have available a driver program to supervise and translate between the language of the data acquisition unit and that of the computer. Programs are available commercially for computer systems that are most commonly used for this purpose. Among these are the RT-11, RSX-11M, and VMS, operating on LSI-11™, PDP-11™ and VAX™ DEC® machines; and PC-DOS,™ used with the IBM PC™ personal computer. Some companies, in particular Hewlett-Packard, supply data acquisition and computing equipment as a package.

6-6 Strip Chart Recorders

Multipoint strip chart recorders have long been the standard means for accumulating records of plant data. They deserve some mention in this text because microprocessor applications have enhanced them so that they have become better able to accomplish their particular uses in industry. The strip chart recorder is a large case instrument, usually front-of-panel mounted, that prints a record of a number of real-time values of process variables on a paper chart that moves past a print wheel. The position of the print wheel in relation to a scale across the top of the chart provides an indication of the variable value. Input signals from process variable sensors are scanned, and the print wheel is positioned for each measurement by a balancing circuit that compares the measured value against a reference. After the mechanism has come to balance, the chart is marked with the point number. The head then indexes to its next point and rebalances. All this time a motor is scrolling the chart, and, as it unrolls and is rerolled on a take-up roll, a trend record is produced on the chart for each of the points.

A special-purpose digital computer built into the strip chart recorder has increased the utility of this device. A small programing keyboard built into the device (see Figure 6-2) allows many variations of measurement not available in analog circuit recorders. In the following comparison of actions that are possible for recorders that use digital circuitry, the word "standard" is used to refer to the strip chart recorder design before the use of microprocessors. List items 8, 9, and 10 are unique to the digital circuitry type.

1. Standard multipoint recorders can accept 4, 6, 8, 12, or 16 inputs. The recorder with digital circuitry accepts 1 to 30 points.

2. Standard recorders can have up to three scales, and indication is achieved by reading the position of a pointer (which moves with the print head) in relation to the scale graduations. The digital circuit recorder displays the value of every point as it is read on a large digit LED indicator.

3. Multipoint strip chart recorders are used extensively for temperature measurement. Transmitted measurements from analog sensors, (thermocouples, resistance thermometers,

and radiation detectors in particular) are not linear with temperature, but will be recorded on a chart with linear graduations unless specially printed (and very expensive) chart paper is used. For standard recorders, a linear indication of input values is achieved by nonlinear spacing of the divisions of the scale. Digital recorders can use evenly graduated scales and chart paper, because lookup tables in memory allow all inputs to be linearized in software.

4. Strip chart recorders can measure as many as three ranges of inputs. The digital circuitry accommodates eight or more types of devices (thermocouples, for example) and a millivolt signal. Values can be displayed in either Fahrenheit or Centigrade units. High and low range values for each input can be defined separately.

5. For the standard instrument, the rate of printing is fixed for the instrument. The digital circuitry allows a selection of print rates from 1 to 180 seconds per point in increments of 1 second.

6. Standard recorders accommodate a high and a low alarm setting for a specified number of points. The digital circuitry allows up to four alarms per point with four different alarm actions: high, low, high rate, and low rate. Printing speed can be increased when an alarm occurs. Alarms can be combined so that there may be an alarm if one point *or* other points exceed their limits; or if one point *and* other points exceed their limits.

7. All points of the standard recorder are included in the scan whether there is an input connected or not. For the recorder with digital circuitry, points can be skipped if it is not convenient to print them.

8. To keep unauthorized persons from tampering with programmed data, a security code can be established and must be used before changes can be made. Displays can be used at any time, however, to review the programming.

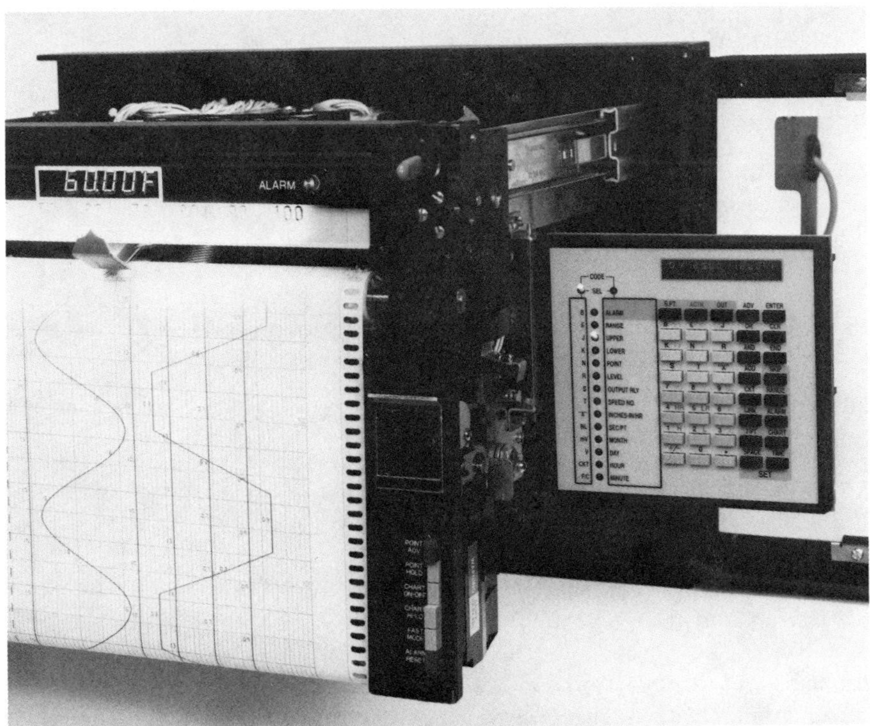

Figure 6-2. Strip Chart Recorder with Programming Panel

(Courtesy Leeds and Northrup Company, a Unit of General Signal)

9. An RS-232C port can be included so that information can be transmitted between recorders or from recorders to devices connected to a communication network.

10. Background diagnostic routines can be run and LED messages displayed to indicate the probable location of a fault.

More sophisticated recording than that available from a standard strip chart recorder can be supplied as part of some data acquisition units. Some of the following records that can be displayed on charts are illustrated in Figure 6-1.

- Analog trending
- Printed reports, formatted to users' requirements
- Graphic reports — bar graphs, histograms, trend records
- Digital record printout in rows and columns
- Event triggered reporting

6-7 Applications

Every user will immediately think of specific applications for data collection. Activities to which data acquisition systems are typically applied, because of the multiplicity of data gathering involved, include:

- Boiler tube temperatures and pressures
- Boiler skin temperatures and pressures
- Turbine generator, pump and fan bearings
- Hydroelectric plants
- Catalytic cracking units
- Tank farms
- Gas and oil pipelines
- Pumping stations
- Drilling platforms
- Test stands
- Collection of data from composition analyzers
- Alarm monitoring
- Energy use computation, energy monitoring, and demand control
- Performance testing

REFERENCES

1. Brown, Jim, "Choosing between a Data Logger (DL) and a Data Acquisition System (DAS)," *Instruments and Control Systems*, September, 1976, pp. 33–38.
2. "Ways to Use Data Acquisition," *Instruments and Control Systems*, November, 1976, pp. 41–45.
3. Kaufman, Edwin M., "Thinking of Changing from Analog to Digital for Data Acquisition?" *Instruments and Control Systems*, July, 1977, pp. 29–31.
4. Person, Terry, and Smed Ruth, "Data Acquisition on a Board: What's Happening?" *Instruments and Control Systems*, March, 1982, pp.53–55.
5. Krigman, Alan, "Data Acquisition Systems: The Microprocessor Revolution," *InTech*, July, 1983, pp. 45–48.
6. Pluhar, Kenneth, "Data Acquisition Diversifies — Introductions Include Loggers, Boards, Systems," *Control Engineering*, October, 1983, pp. 59–62.

EXERCISES

1. What criteria should be considered when selecting a data logger; a data acquisition unit?

2. Library assignment — list manufacturers of data acquisition equipment and compare the listed specifications for quantity of inputs, types of inputs, and other parameters.

3. How does the application of a data acquisition unit influence the type of analog-to-digital conversion method used?

4. How can a data logger be used as part of a plant safety system?

5. How might a data logger be used for optimizing?

UNIT 7
Single-Loop Controllers

The processes to which single-loop control can be applied may be continuous — as in boiler control, where a boiler is fired 24 hours a day, seven days a week, and generates steam continuously; or they may be batch processes, in which a predetermined sequence of operations is performed to produce a batch of product, then repeated to produce another batch. In either case, the controller functions to maintain a specific process variable at a desired value. The desired value corresponds to satisfactory process operation. It is referred to as "set point" or "reference value," and the function of the control loop is to make the measured process value match the set point. This type of control is called feedback control or closed-loop control, because the result of a change in output is "fed back" to the input for comparison with the reference value, closing the loop from input to output. It is the primary application of single-loop controllers. Figure 7-1 illustrates the classical feedback loop, including a process with internal and external load changes. The process is always part of the feedback loop.

This description of control is generic. It can be accomplished by either analog or by digital components. In order to define how the digital implementation differs from its analog counterpart, it will be necessary to go into some detail about the ways that feedback control is accomplished by instrumentation.

7-1 PID Control

Not so many years ago, continuous and batch control applications were performed only by analog devices. The most important of the analog devices used for process control was the single-loop controller. In its basic form it performed feedback control. In other words, the purpose of the controller was to reduce an error signal (deviation of measured value from set point) to zero.

There are three ways in which feedback control uses the deviation of the measured process value from set point to develop output values. They can be demonstrated by making

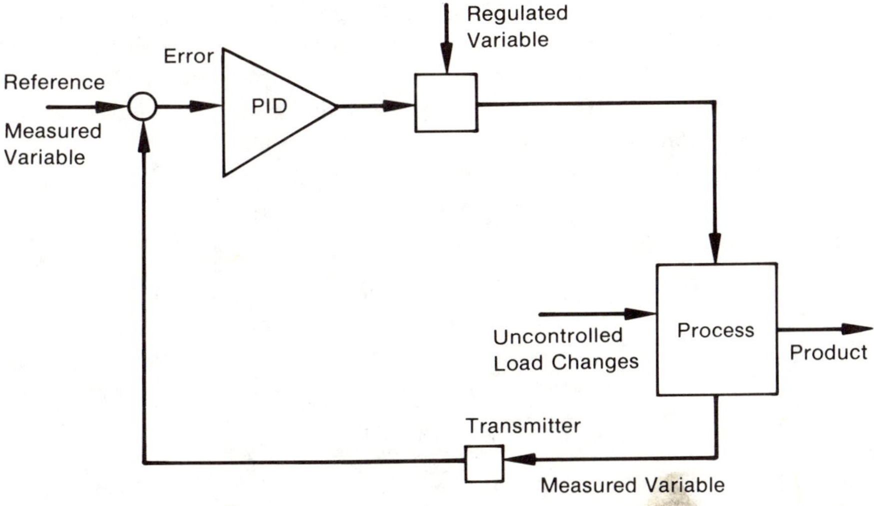

Figure 7-1. Feedback Control Loop

a set point change to a system that is at equilibrium, with no deviation and a specific output value. Making the set point change creates a deviation, and an output change will result.

1. The output value change can be a function of the amount of deviation from set point. This is called "proportional control", because the change in output is proportional to the deviation. A gain factor can be programmed so that the degree of proportionality is selectable; in other words, output change for a given deviation can be larger or smaller, depending on the gain.

2. It can be a function of the length of time that the deviation has existed. If the change in output does not correct the error during a programmable length of time, another correction proportional to the deviation remaining is made. Another way of saying this is that the correction is integrated over the length of time that the deviation exists until zero deviation is finally attained. This mode of control is called "integrating control". Another term used is "reset" because the correction is reset and repeated as long as there is a deviation.

3. It can be a function of the rate of change at which deviation is taking place. A rapid change results in a larger initial correction than does a slow change. This is called "derivative control," rate action, or anticipatory control.

Single-loop controllers have become known as **PID** controllers because they can be adjusted to produce responses to deviations that are functions of a combination of **P**roportional action, **I**ntegral action, and **D**erivative action. The set point value of the analog controller is selected by adjusting a knob on the front of the housing until a pointer on a dial indicates the desired value. The plant operator can make this adjustment. The PID function values are also selectable by positioning potentiometers, but these adjustments are located inside the housing and are usually changed only by an instrument technician, an engineer, or a supervisor.

7-2 Controller Housings

The single-loop controller is usually housed in a front-of-panel mounted enclosure, with the displays of process variable value and control output value visible to the operator. Manual/auto selection, set point adjustment, and manual output adjustment are also available to the operator from the front of the panel.

Another housing design known as "split architecture" divides the controller hardware between two housings. One housing, mounted on the panel, lets the operator see a meter that shows process variable value, another meter that shows output signal percent, a knob and dial for set point adjustment, and a switch for selecting between manual and automatic control. A second housing, located behind the panel, or even in another room, connected by cable to the panel-mounted portion contains the electronics that perform the error detection and control functions. The PID adjustments are in this part of the device and are not accessible to the operator at all. This type of analog control has been very popular in steam power utility installations and chemical plants.

7-3 Controller Adjustments

All analog control uses the P (proportional) mode of control, and almost all includes the I (integral) mode. This combination is known as PI control. Most designs include the D (derivative) mode also, providing a full-fledged PID capability. To a degree, the PID adjustments model the process to which the controller is applied and allow adjustment of the response characteristics of the controller to match the process lags and capacities that prevent instantaneous process response to corrective action. Other factors affecting controllability of a process are interactions between processes, instability created when more than one variable changes to affect the final measured variable value; and upsets outside the loop, caused by supply and demand changes but creating changes in process characteristics.

Functions more sophisticated than PI and D have been developed to accommodate these additional influences. They have been given names like cascade, ratio, feedforward, adaptive tuning, floating control, and programmed set point. Sometimes the components performing these additional functions can be included in the same housing as the PID control, but often more than one housing is required, with hard-wiring interconnecting the various boxes. They are sufficiently important to an understanding of process control to warrant further explanation.

7-4 Ratio Control

Sometimes more than one variable must be changed in a controlled system to properly maintain process conditions. A familiar example is combustion control, where a relationship between the quantities of fuel and air must be maintained, regardless of the amount of fuel needed to produce energy. Another example is the preparation of gas mixtures. When combinations of hydrogen and nitrogen, for example, are produced as atmospheres for heat-treating furnaces to prevent oxidation of the metal being heated, the relationship between the gas components must be exact. Figure 7-2 shows a generalized scheme for performing ratio control with analog equipment. Flow of one component is measured and controlled directly. Flow of a second component is measured and controlled by a loop whose set point is a ratio of the measured flows of the first and second components. Because other parameters can affect the measurements (gas compositions are a function of pressure as well as quantity, for instance), it is recommended that the composition of the mixture be measured and the ratio reset if the final value is not correct, using a separate control loop.

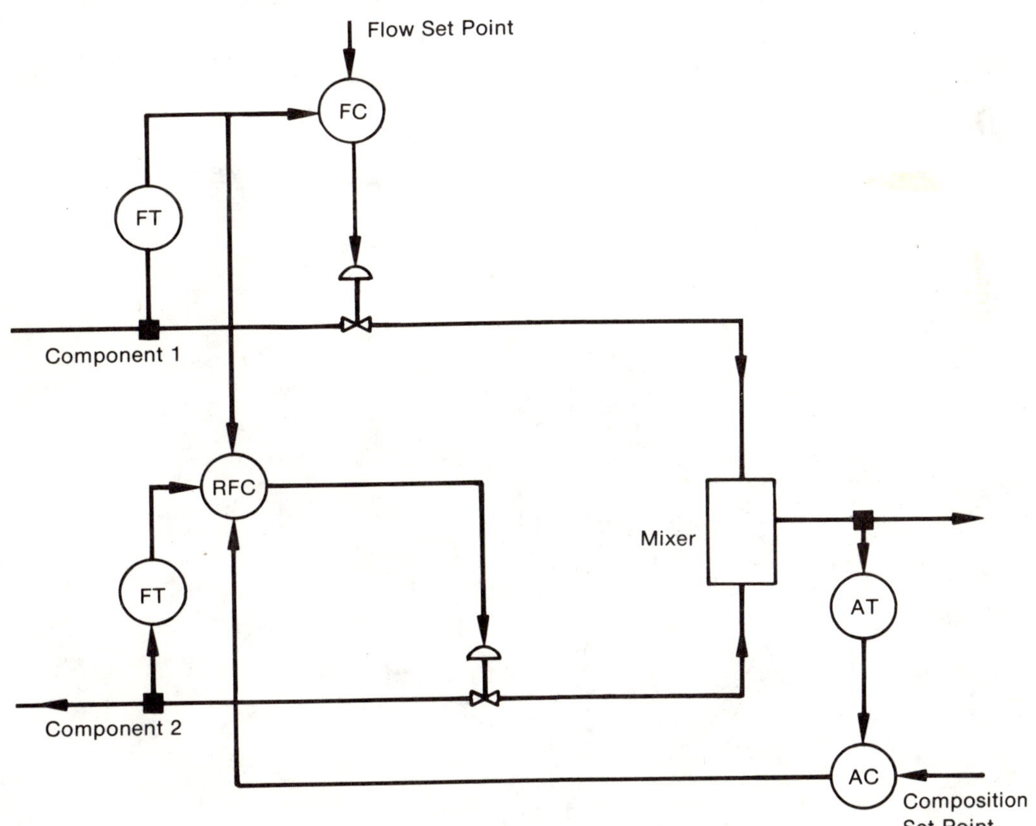

Figure 7-2. Ratio Control

7-5 Cascade Control

Process control may be affected by variables outside the process itself, for example, by changes to the utilities that provide energy to the process. This can happen in a heating process using a heat exchanger, where oil is heated by steam. The heat in the steam will vary as the steam pressure varies. Single-loop feedback control, shown in Figure 7-3(a) will measure the temperature of the heated oil leaving the coils of the exchanger and will vary the amount of steam to the jacket to maintain that temperature at set point. If the steam supply is used in other parts of the plant, however, its pressure may vary because the boiler supplying it cannot instantaneously meet all the demand; therefore, the amount of heat corresponding to a given steam control valve position is not an absolute value. This type of change is called a demand load change. The controller will detect the fact that the measured temperature is not exactly at set point and will make corrections, but there is a significant time lag between the time temperature is sensed and the time when corrected energy flow has made its way through the physical barriers separating steam from oil in the exchanger. The control valve will always be trying to catch up with changes, and control will not be as good as it should be.

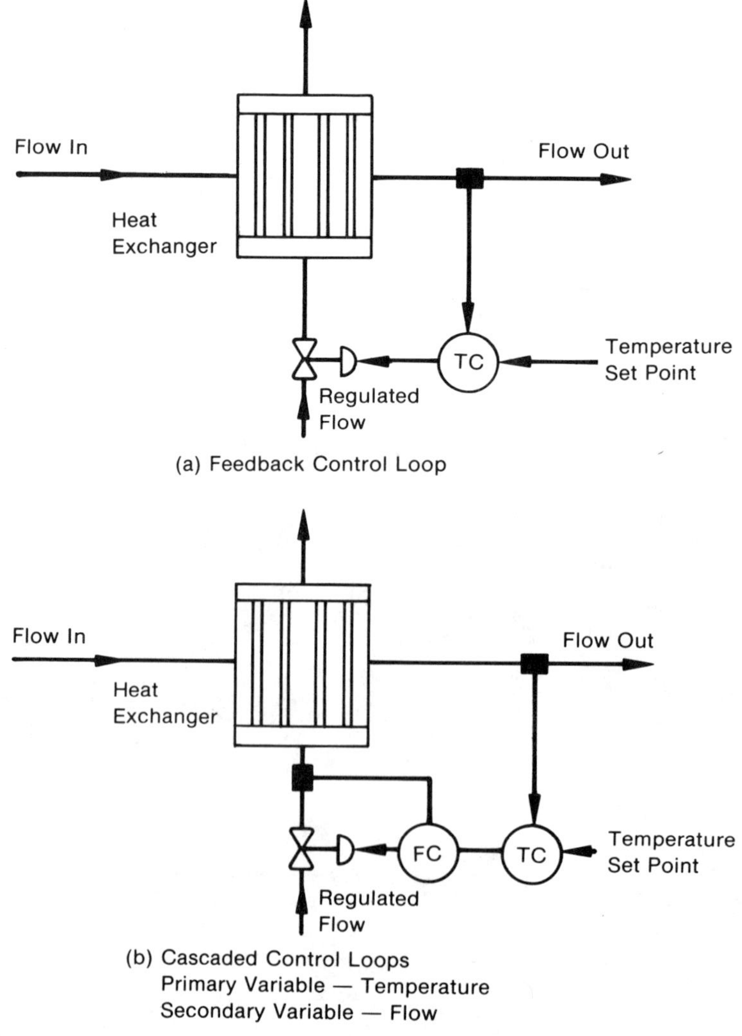

(a) Feedback Control Loop

(b) Cascaded Control Loops
Primary Variable — Temperature
Secondary Variable — Flow

Figure 7-3. Cascade Control

Temperature control loops are usually slow, for the reasons indicated above, but flow and pressure loops are fast, because these variables respond immediately to changes in control valve positions. If the flow of steam could be continuously adjusted to a value that would always provide the right amount of energy to satify the requirements of the measured temperature, valve position would no longer be a factor. This can be done by cascading the output from the temperature controller (using it as a set point value) into another control loop that measures steam flow as its process variable. This is illustrated in Figure 7-3(b). The deviation from set point of the temperature controller (the primary controller) creates an output that is used as the set point for the second controller. This set point will not change when the temperature (primary) controller set point is satisfied. If it is not satisfied, because of an oil temperature or flow change, or because the set point is adjusted to a new value, reset will make the primary controller output change until there is no deviation between set point and the measured temperature. As a consequence the set point of the secondary controller will change, and a proportional change in secondary controller output will result in a change in the energy flow to the exchanger. Changes in temperature because of steam energy changes will be detected by the output temperature measurement, and correction by changing the steam flow will be rapid, so that the outlet temperature of the exchanger is controlled to a much closer degree than by the single loop of Figure 7-3(a).

Parenthetically, some digital control systems use the expression "cascade mode" to mean that the set point is developed outside the controller. There will, therefore, be a difference between "operation in the cascade mode" and "cascade operation". Cascade operation utilizes a primary controller, with its output becoming the set point to a secondary control loop. Operating in cascade mode can mean that the reference value of a controller is generated by the output of a primary controller, by a computer, or by any remotely located device that has an analog output of the proper span.

To complete the definition of "mode": whenever the set point is developed internal to the controller itself, the controller operates in automatic mode. There is a third mode of operation, manual mode, where the output is generated manually by the operator.

7-6 Feedforward Control

Cascade control can often be used advantageously to correct for external load changes, but it will not be effective if deviations are caused by process changes. Sometimes more than one process variable changes. Figure 7-4 shows a control system applied to a waste treatment facility where acid waste is neutralized in a very large mixing vessel by adding alkaline solution. If the influent flow rate changes, or the acidity of the influent changes, the acidity in the tank will change but the pH measuring device will not detect the change immediately. When it does do so, alkaline reagent will be added to correct the problem. The neutralization takes place rapidly, but, because the mixing vessel is very large, a long time will elapse between the addition of the neutralizing solution and the measurement of its effectiveness. The controlled variable will over-correct and drift from one extreme to the other around set point. One solution is to measure the flow rate of the acid waste and also to measure its pH; compute the amount of reagent that should be added to correct for the flow entering the tank; and feed this forward as set point to a controller that adjusts the amount of reagent addition. Corrections are made immediately for demand changes — changes in acid waste load or acidity. Feedback control is often combined with feedforward control to trim the computed set point if final results indicate that other changes are having an effect on the control process.

This system appears to be similar to that used to illustrate ratio control, Figure 7-2, but in this case the process is very slow, and the controller will be a much less active component in the loop. Multiplication and function generation and summation of measured variables are also involved, effectively modeling the process to compute the anticipated set point. Process modeling can be done by analog devices, but a number of them must be combined into a hard-wired system, carefully designed, to accomplish the requirement.

7-7 Programmed Set Point Values

Some batch processes operate with one set of parameters during part of the operation but require other parameters during other parts. A heat-treating process, for example, may consist of a series of intervals of time when the material being treated soaks at different temperatures, with ramps at continuous rates from one soak level to the next. Figure 7-5 illustrates such a time-temperature relationship. The material to be heated is held at 10% of the temperature range for an hour; heated at a rate of 20% per hour to the 30% level; soaked at this level for 2½ hours; heated at a rate of 11.11% of range per hour for 4½ hours, to the 80% level; soaked for 2 hours; and then cooled at a controlled rate for one hour to the 0% point. This type of control action has been achieved for many years by combinations of feedback control loops, with timers, integrators, and relays accomplishing the changes in set point value.

7-8 Adaptive Tuning

A long-sought control function was finally realized commercially a few years ago by analog control manufacturers after being talked about for many years. This function is adaptive control, a means of varying the proportional band adjustment (and in some cases the integral adjustment, as well) to accommodate variable process conditions. Programmed set point operations provide an example. Response times and process lags, the basis for determining tuning constants, may be very different at the beginning, during the main portion, and at the end of a batch process. When the ramp and soak set point adjustment described in the preceding section is used, changing a control valve position ten percent because of a ten percent change in deviation between measured temperature and set point may produce an entirely different effect when cold material is being heated from the effect of a ten percent change when the material is hot. If proportional band is not changed, that is exactly the way the valve will behave. It might be better, under certain conditions, to make a one percent change in valve position for a ten percent change in deviation, but until recently controllers had no way of knowing that. Electronic circuits have been developed, however, that can continuously experiment with the results of output changes and can revise proportional band and reset settings until the best ones are found to match the current process conditions.

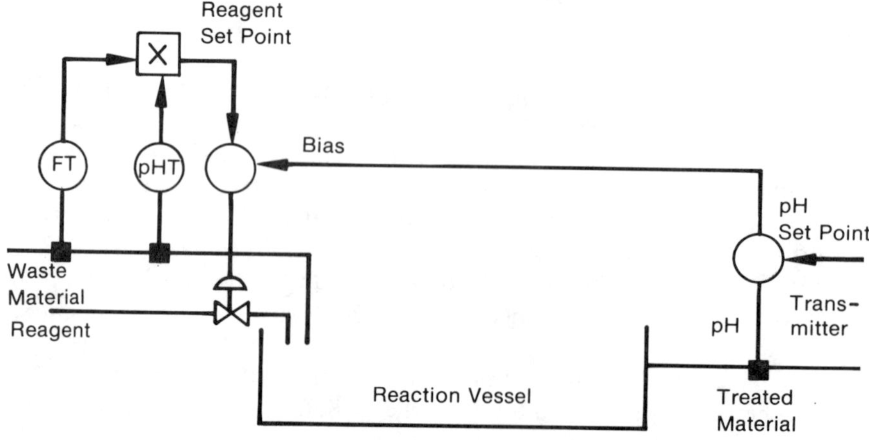

Figure 7-4. Feedforward Control with Feedback Trim

Still other functions are accomplished by analog devices. These included creating alarm indications to announce to an operator the occurrence of a process condition outside safe or desirable limits. They include the computation of values: taking the square root of a measured variable, adding several values together, multiplying and dividing values, integrating values to show usage over a period of time, and amplifying the outputs of sensors that develop small signals to levels that can be comfortably handled by the analog device. They include devices for linearizing signals from sensors whose outputs are not linear with respect to the process variable values. They include the generation of variable set points to accommodate changing conditions during batch operation.

All of these functions — control, computation, alarming — can be done with analog devices. In most cases, a separate piece of hardware is required for each function. A system of functions requires a lot of panel space and much interconnecting wiring. There are other limitations. Many single-loop controllers are designed to perform only the most elementary tasks; temperature control is the most common application. Temperature controlled processes usually change slowly and can be kept close to set point by using uncomplicated tuning parameters. Single-loop designs can be turned out in large quantities for such an application.

Other types of measurement demand more sophisticated controller tuning. Flow and pressure, the other most commonly used variables, are usually harder to control than temperature because they respond to load changes very quickly; pH, a variable encountered in waste treatment processes, is nonlinear. A further complication of pH control is that abnormally large proportional band settings are required because the process includes large time constants, and valves should move only small amounts for normal process changes. Usually, there are fewer flow and pH control loops than temperature loops in processes, and it is uneconomical to design special single-loop controllers to take care of them.

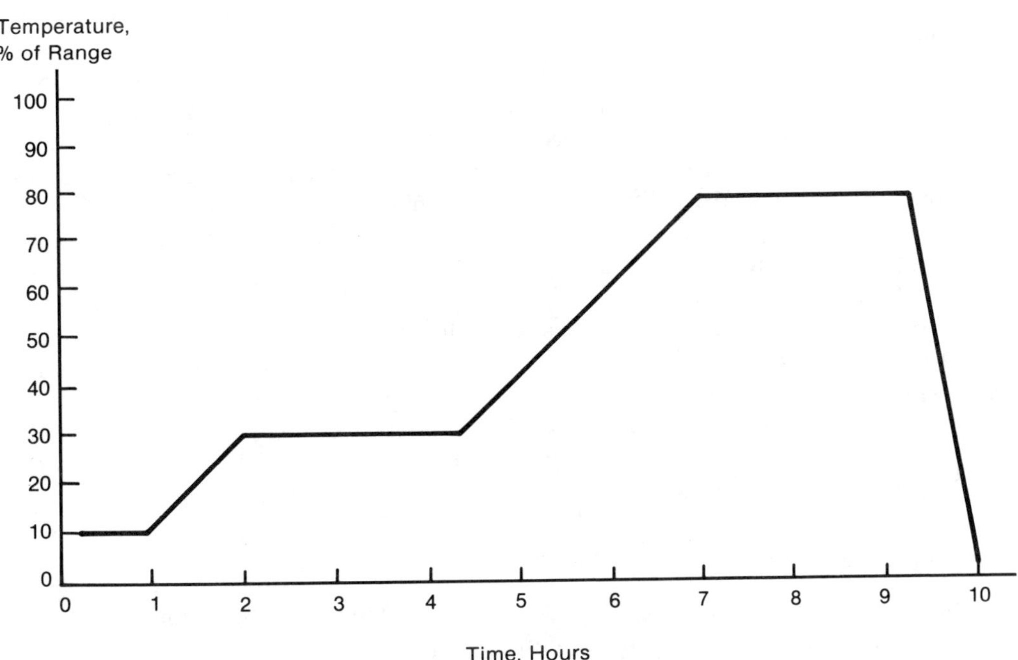

Figure 7-5. Ramp/Soak Programmed Set Point

7-9 Digital Single-Loop Control

The reason for spending so much time describing process control is to establish a background of application that will make clear the role that control instrumentation plays in industrial process. A subsidiary reason is to demonstrate that the type of applications that require the combination of a number of analog devices may require only a single digital device. All of the functions described above can be accomplished by digital control, but without requiring a multiplicity of hardware and interwiring. With the advent of the microprocessor and the reduction of the price of memory it has become practical to build a small special-purpose computer into a single-loop controller, making it much more powerful and flexible than the name implies. Today's single-loop controllers are still packaged in a single housing, but programming permits combining a number of the functions listed above so that their performance can be easily matched to the characteristics of the process and load conditions to which they must interface. In addition, some single-loop digital controllers come with communication ports, permitting connection to highways, personal computers, and networks including other controllers.

Single-loop control is implemented digitally in at least three different ways. The first is a direct manifestation of the name, a housing providing the functions of a single loop of feedback control. The second is sometimes called multi-loop control, but physically it consists of a single housing with several (usually four) individual loops of feedback control. The electronics are time-shared. The functions of the loops may be interconnected by the user, but they can also be used as individual loops with inputs and outputs completely independent of each other. A shared LED display and push button keyboard allows the operator to observe process conditions and make tuning adjustments.

7-10 Special-Purpose Controllers

The third manifestation is the special-purpose controller, a preprogrammed interconnection of several (again, four is a common number) loops that act together for a single purpose. One common example is combustion control. The four loops may provide control of fuel flow, control of air flow at a ratio to fuel flow, a bias to the ratio generated by deviation from set point of the measurement of oxygen in the stack gases, and furnace pressure control. Another combination that has been marketed is specific to the control of compressors in gas pipe lines. The controls in this case may be for gas flow, outlet pressure, and surge control based on pressure ratio, compressor speed, and bearing temperature. Still other special-purpose computers have been developed for monitoring purchased electric power and shedding power loads so that the utility's consumption limit is not exceeded.

This type of small system control can be successfully marketed if there are enough duplicate applications to develop a market forecast that will support the cost of developing the package. It can only be done cost effectively by using the programming capabilities of microprocessor-based digital circuitry. In evaluating its cost, the costs of specific sensors must be included because, unlike the single-loop controllers of the first two categories, which can be applied for use with almost any process variable transmitter, each supplier of special-purpose controllers will include his own sensors (oxygen measuring probes, for example) as part of the package. There is usually, however, one principal controlled variable, and so the system qualifies as a single-variable, if not a single-loop, controller.

7-11 Algorithms Provide Special Functions

Most measurements are inferential; that is, they infer the real process condition, which cannot be directly measured. It would be desirable to control some batch chemical reactions on the basis of completion of reaction, but there are few sensors that can measure composition in real time (the measurement can be determined in a laboratory but becomes available far too late to be used for controlling the process). The next best thing is to measure temperature or pressure and infer that the reaction is progressing satisfactorily.

Since any digital controller is a computer, it is possible to include a library of functions in memory. Once developed, they can be combined by programming to compute new, composite functions. The library-based functions are sometimes called algorithms; the PID function is an algorithm on a chip. Some libraries contain many functions, and they can be combined by programming. This adds a powerful dimension to the single-loop computer capabilities. Examples of the usefulness of this programming capability are included in Unit 12.

An extension whose time has not yet come is the microprocessor-based unit with algorithms that directly take into account the multivariable characteristics of unit processes. In some processes more than one variable has a significant effect on the process condition, and there is a continuously adjustable optimum combination that will produce the maximum quality and profit at the minimum cost. The output of a fan, for example, may be affected by speed as well as the position of dampers, and both variables can contribute to an optimum condition.

The mathematics of multivariable interaction are known, as are the interactions of specific chemical processes. Paper and pulp production, chemical reaction, fractionation and distillation, and mechanical reduction and separation are typical examples. When these can be produced as menu-selected algorithms subject to programming to fit them to specific actual applications, closed-loop control will reach a potential that far exceeds its present capability.

Figure 7-6. Single-Loop Digital Controller

(Courtesy Fischer and Porter Company)

7-12 The Digital PID Controller — Hardware Design

From the front of an instrument panel a typical single-loop controller, digitally implemented, does not look very different from its analog counterpart (see Figure 7-6). The same operator interface is necessary, so there will be indications of process variable and of output control signal value; means for selecting manual or automatic control mode; and adjustments for set point increase and decrease and for output signal increase and decrease. The indicating devices may be gas discharge tubes and liquid crystal displays instead of millivolt meters, and the adjustments may be made by pressure-sensitive membrane touch pads instead of by toggle switches, potentiometer knobs, and push buttons. Gas discharge tubes and pressure-sensitive touch pads are appearing in the design of analog circuit controllers as well, since they have demonstrated pattern-recognition related benefits that make an operator's job more productive. Other important hardware differences are the use of plug-in printed circuit cards with flat, flexible cable connections. Using plug-in cards simplifies manufacturing, spare parts support, and field changes. The important software differences are the programming capabilities that allow all of the above mentioned functions to become available in one assembly.

A typical digital single-loop controller will have several printed circuit assemblies mounted on a backplane or interconnected by cables. These will provide the functions of power supply, input and output processing, and computation requirements. There will be a microprocessor, which will coordinate the acquisition of data, the calculation and manipulation of data, and the transfer of data between printed circuits within the instrument; will control the output signals; and will provide communication to external devices through output ports. It does all of this by combining input data with operating constants held in RAM according to a programmed relationship maintained in ROM.

The processor performs the following activities under control of the program stored in ROM:

(1) Scan the analog input terminations and digitize the analog values, then store the binary bits produced in RAM.

(2) Scan the digital input terminations and store the bit arrangements in RAM.

(3) Scan the information input from the operating keyboard.

(4) Based on the stored information, update output values and store them in registers addressable to output ports.

(5) Set analog output levels by loading data into the proper D/A converter and strobing proper sample and hold circuits.

(6) Control the digital displays on the front panel.

(7) Run diagnostics and reset watchdog timers unless a failure has occurred.

(8) Handle serial data communications with host computers, printers, and other external devices. Format serial data if necessary.

A number of regulated power supplies will be required. There will be one intermediate voltage (approximately 20 to 26 volts dc) supply. This may be a switching supply, which regulates the switching of a transformer primary through the use of MOSFET switches. Regulated power at lower voltages will be drawn from the intermediate voltage supply. Popular regulated voltages are ± 5 volts dc, ± 12 volts dc, and ± 15 volts dc.

7-13 Networks

The eighth characteristic listed above is an example of the overlapping of categories of application encountered in discussing digital logic control. The single- (or four-) loop control package will often include, at least as an option, a communication port through which communication with other devices can be established, directly or over a data highway. This allows the single controller to become part of a global system. It can create an alternate to the use of distributed control systems, performing many of the same functions. An advertised benefit is that the user can get all the advantages of CRT-console control without giving up the familiarity of panel-mounted instruments. At the same time it will not be necessary to invest in panel space and hardware for annunciators, rack-mounted auxiliaries, recorders, and large numbers of wired connections. Several suppliers of small panel-mounted controllers offer the capability of including several hundred single-loop controllers in a single data highway connected installation. The interested purchaser must investigate the pros and cons of the alternates very carefully, because the time required to transmit information from this many devices may have a significant effect on the controllability of some processes.

REFERENCES

1. Miller, Timothy J., "Supervisory Unit Links Controllers to Form Distributed Control System," *Control Engineering*, October, 1983, p. 93.
2. Bailey, S. J., "'Soft' Temperature Controls Give Users Exact Loop Fit with Standard Hardware," *Control Engineering*, October, 1983, pp. 83–88.
3. Andriev, Nikita, "A Process Controller that Adapts to Signal and Process Conditions," *Control Engineering*, December, 1977, pp. 38–40. 4. Hausman, Jack F., "Applying Adaptive Gain Controllers," *Instruments and Control Systems*, April, 1979, pp. 51–55.
5. Yarber, William H. II, "Single Loop, Self-Tuning Algorithm — Applied," Session #31, "Software for Advance Control, 1984 AIChE Anaheim Symposium.
6. Murrill, Paul W., *Fundamentals of Process Control Theory*, Research Triangle Park, NC: Instrument Society of America, (1981).
7. Shinskey, F. G., *Controlling Multivariable Processes*, Research Triangle Park, NC: Instrument Society of America, (1981), pp. 117–175
8. *pH and pIon Control in Process and Waste Streams*, New York: John J. Wiley and Sons, (1973).

EXERCISES

1. What is "dead time" and how does it affect controllability?

2. What is reset limiting and how does it affect batch control?

3. What is meant by decoupling?

4. Explain floating control and error-squared control.

5. The output of a PID controller can be configured to increase or to decrease, as the process variable becomes greater than set point. An increasing output, under these circumstances, is called direct-acting control; a decreasing output is called reverse acting control. Which one should be selected?

UNIT 8
Programmable Logic Controllers

Programmable controllers are microprocessor-based special-purpose computers that can be programmed to perform functions that replace relays and electromechanical stepping devices. At least, that was the intent of their design when they were first introduced during the 1970's. They used the acronym PLC, for Programmable Logic Controller. Since then there have been extensions in design to include the capability of PID (proportional, integral, derivative) control to perform closed-loop (feedback control) applications, and the marketing departments of their producers feel that the logic capability should be made less of while emphasizing the analog control capability. Consequently, the acronym PC has become common, signifying that these are truly Programmable Controllers. Unfortunately, IBM has made the same acronym popular for its personal computer so there is sometimes confusion, especially if you happen to be the owner of an IBM PC™ personal computer. Alternates have been suggested, including Data PC, and PμC (pronounced "puck"). In this discussion, we will refer to programmable logic controllers as PLC's, and the reader may substitute any other personal choice of term.

8-1 Functions of the PLC

The basic task of the programmable logic controller is to accomplish electronically the functions performed by electromechanical devices such as relays, counters, timers, drum sequencers, and stepping motors. In the form in which it was originally introduced to the market, it accepted discrete (on-off) signals from externally located devices with contacts that close to pass ac or dc voltage signals; and it created output conditions that caused electrical voltages to actuate relay coils, solenoids, lights, annunciators, and other electrical loads. PLCs still perform this action very well.

The expanded form of PLC available on the market today can accept continuously variable analog signals from process variable measuring devices and from transducers of these signals; and it can develop outputs that are continously variable analog signals. These can be used to operate regulators such as current-to-pneumatic converters and silicon-controlled rectifiers. Some PLC's can accept low level digital inputs, pulses, and millivoltages directly from process measuring devices such as thermocouples and strain gages.

The designs first produced and sold were intended for use in machine tool control applications. Extended designs available today can also be used for industrial process control applications. The majority of the sales of PLC's are still for machine tool, material handling, and mechanical equipment sequence control applications, but the number of process control application sales is increasing. PLC's are very competitive with other digital control devices more usually associated with process control applications, particularly distributed control systems and systems using microcomputers as the controller. The more expensive competitive systems provide computational and graphic display capabilities, areas in which the PLC has been limited. It is, however, catching up rapidly.

As new developments reach the marketplace, the appearance of the PLC is changing. When programmable logic controllers first appeared at trade shows the displays featured relatively small modules, sometimes rack-mounted, with off-line programming assemblies. These had the only man-machine interface, a table model video screen display. The CRT's were small, usually having screens with a diagonal dimension no longer than 12 inches, and had a black and white display. Over the years, colored displays with process graphics began to appear in the PLC booths. Today almost all PLC suppliers can provide consoles and large video screens, a significant advance for process control applications. The PLC design, however, still emphasizes logic control and sequential operation.

8-2 Hardware Description

Three separate functions are necessary to make up a PLC installation. They are provided by the power supply, the processor, and the input/output hardware. Typical units are pictured in Figure 8-1. Peripheral devices, including display monitors, programming devices with keyboards, and printers, may be added. For systems with only a small number of inputs and outputs, all three components may be packaged together. More often, each of the three necessary functions is packaged separately and may be mounted in separate locations. The power supply and the processor are often located in a motor control center, or some central area, while the input and output modules may be local to the processing areas. Many models can communicate over a data highway, and a minicomputer may be interfaced to the system. PLC's are programmed off-line, and their operation is completely transparent to operating personnel.

Power Supplies

Power supplies convert 120 V/240 V ac power to the dc voltage necessary for operation, usually 5 volt and 12 volt dc. A power supply will provide enough power for the processor and for a number of inputs and outputs. The load will vary, depending on the number of inputs and outputs that are actuated, but a power supply should be sized to handle a maximum condition. The measures recommended for power supply support equipment in Unit 4, particularly for constant voltage transformers to adjust supply power subject to variations and isolation transformers to protect from electromagnetic interference, should be observed.

Input and Output Modules

Input and output circuits may be combined in the same module or they may be separate, depending on the supplier. An individual module will have a number of terminations for circuits, usually in groups of six or eight. The terminals may be mounted on an arm that can

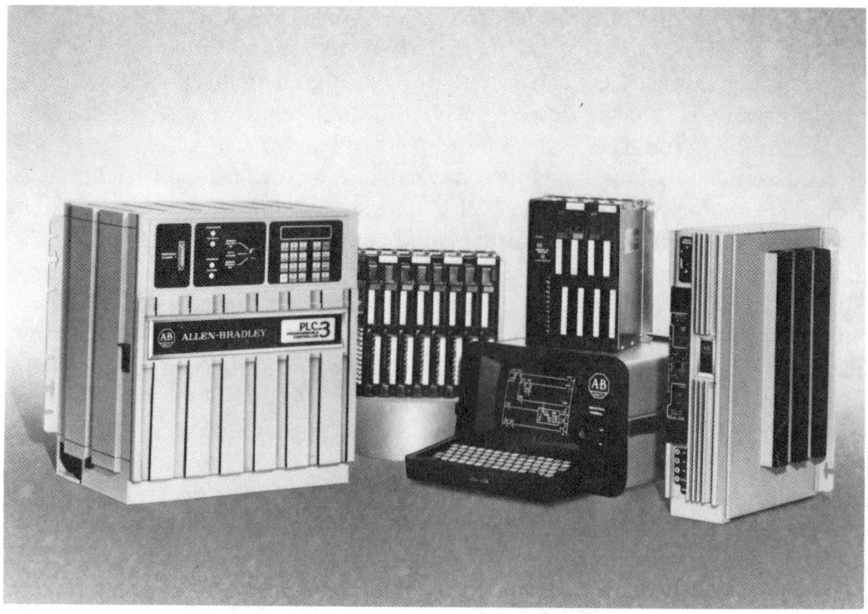

Figure 8-1. Programmable Logic Controller Components

(Courtesy Allen-Bradley Company)

Table 8-1
Input and Output Signals — Programmable Logic Control

Inputs Source	Input Type	Remarks
Limit switches	120 V ac/125 V dc	Input filtering
Float switches		
Selector switches		
Push buttons		
Proximity switches	120 V ac, 5 to 30 V dc	
Photoelectric switches	12, 24, 48 V dc	
TTL signals	5 V dc	
Encoders	5, 12, or 24 V dc	Input pulses can be counted. Rates up to 50K Hz.
Quadrature type		
Incremental type		
Optical beam counters		
Eight-bit Gray code with zero speed switch	12 to 24 V dc	Signals from absolute encoders
Thumb wheel switches	12 to 24 V dc	
ASCII code	12 to 24 V dc	Interface signals to peripheral devices
Thermocouples, RTD's	Millivoltages	Temperature measuring devices

Outputs Load	Voltage level	Remarks
Relays	120 V ac, 2.0 amps	Triac output
Solenoids	220/240 V ac	
Motor Starters		
TTL circuits	5 V dc	
Logic Circuits		Reed Relays
Switching relays	0 to 250 V ac	Mercury wetted relays
Motor starters	0 to 175 V dc	
Variable Speed drives		
Servos	±10 V dc	Axis motion control for grinders and material handling

Input/Output		
Analog circuits	dc; 0 to 5 V, O to 10 V, 4 to 20 mA	Usually 2 per module

Miscellaneous		
Clocked pulses		Used by the processor for calendar and time comparison

be swung away from the module, making the connection of wires easier than if they must be connected directly to the module. Input circuits may have 1500 rms isolation.

Output circuits are usually fused and may be equipped with lights to indicate when a circuit is energized. There may be one blown fuse indicator for the entire module, or separate indicators for each circuit. Input circuits are often equipped with lights to show operational status. Units can be mounted individually or in groups in relay-rack type housings.

There are many variations of input and output signals that can be connected to PLC's. They are summarized in Table 8-1. Special I/O modules that function independently of the processor may be required for use with high speed inputs. These include latching input modules for retaining high speed pulses, and high speed counter/encoder input modules for accumulating pulse inputs from pulse trains.

For best results attention must be paid to installation wiring. The manufacturer's recommendations should be carefully observed. As a general rule, grounding is important to avoid ground loops and the effects of electrical noise. The path to ground from all components must be permanent and continuous. It is recommended that the components in any enclosure be connected to a ground bus and from there to ground, using 8 AWG or larger copper wire or 1-inch copper braid. Each connection should be run separately to the ground bus and from there to earth ground.

Wiring from discrete signal and load devices can be single conductor 12 or 14 AWG. Wiring from low level dc signal producing devices should be shielded, twisted pairs, with the shield grounded at one end only. Devices in this category include TTL signal links, encoder signals, servos, 4 to 20 mA signals from process variable transmitters, and dc proximity switches.

Data carrying signal communication cables should never be run in the same ducts as high power wiring.

Processors

The processor is the part of the PLC that qualifies for the term "computer". It contains the memory, the microprocessors that control the executive and operating programs, and registers where computations take place. The activities of the operating programs include data handling between the processor and the input and output devices, the mathematical and logical operations performed in accordance with the programs written for the device, and diagnostic routines. The size of the programmer is a function of its memory, and this in turn is a function of the number of inputs and outputs. Small units may handle as few as 8 to 32 inputs and outputs, with 2K bytes of memory. Large units can work with 4096 inputs and outputs, and memory size as high as 240K bytes is advertised. For even larger system requirements, several processors can be combined through communication networks and programmed to communicate with each other, with video displays, and with peripheral computers.

During program execution, all the inputs associated with a processor are scanned and brought into the working memory. PLC scan time is generally the time to execute the instructions in a user program one at a time. However, published scan time may include only the time it takes to scan the logic functions, or it may include the time it takes to do internal diagnostics, update the input and output values, and communicate with peripherals. Then again, these may be done every scan or only on an interrupt basis. Interrupts allow the processor to jump to and execute internal program logic or subroutines. As an added complication, certain logical operations may be scanned on a higher priority basis than others.

It might be thought that the architecture of the microprocessor would have a bearing on scan time. This is not true, however. A 16-bit machine will certainly function faster than an 8-bit machine — but if it has more tasks to do, its scan time will be longer.

Memory can be thought of as an array of rows and columns. Computation is performed by pulling data from memory and putting it into a special location in RAM, processing it, and returning computed values to memory. The control logic dictated by the programming is executed, and output values are computed. Finally, the outputs are updated. If only discrete inputs and outputs are involved, the entire process will take place in a few milliseconds. The actual number depends on the size of the system, but a 100 msec scan is a long time for this type of installation. However, scan times must be fast enough so that rapid information changes, as might occur in machine tool applications and in high speed counting, are accounted for.

Scan rates for large controllers as fast as 0.15 msec/K of ladder logic are advertised. Two to 5 milliseconds/K is a more commonly encountered figure and is considered to be very fast.

If analog inputs and outputs are scanned and PID functions are performed, scan times are much extended. As much as 500 milliseconds may be required for a few loops of control. If there is a mixture of analog control and digital logic in the same system, consideration should be given to operating each type from a separate processor.

8-3 Programming

A PLC is a special-purpose computer, and it must be programmed to accomplish its specific special purpose. A processor may come with a touch pad keyboard and an LED display built into it so that programming information can be entered manually, but it is more common to program off-line, using a special keyboard in conjunction with a video display screen. Personal computers are being used increasingly for programming PLC's.

A programming language should allow the user to communicate in a normal manner. This is why BASIC is commonly used as an introduction to high level languages; it appears to be modelled on the English language. Actually, it is much more complicated, and as used by professional programmers who understand its structures and can make use of them for sophisticated programming, it does not read like English at all. Nevertheless, it is a good introduction to programming for the uninitiated.

In the same way, programming for PLC's is most frequently done by configuring a graphic picture on a CRT screen. To promote normal communication the picture looks very much like the relay ladder diagram that is used by an electrician for hard-wiring contacts and coils of relays. This is a kind of language that design engineers and technicians who have worked with relay logic are accustomed to reading and using. Its diagrams are familiar, they offer real-time monitoring to indicate what is happening in the circuits, and they can be documented by computer programming. Documentation may simply be a printout of the ASCII equivalent of the program entered, or, using programs that are not usually standard with PLC's, it is possible to print out the ladder diagram with mnemonics to indicate the tags of the various devices in it.

Programming may also be done on the basis of using Boolean logic. To people who use it frequently this is an equally understandable language. Interestingly enough, engineers and technicians who have spent many years working with relays and ladder diagrams are not always familiar with Boolean logic; students who have learned logic design using Boolean algebra may never have worked from a ladder diagram and will have no idea what it is. Accordingly, some description of both forms of display will be given.

8-4 Ladder Diagrams

"Ladder diagram" is the name given to a schematic wiring drawing that shows its elements on successive lines, like the rungs of a ladder.

Figure 8-2 is a typical ladder diagram representation of schematic wiring. The power and neutral buses form the side rails of the ladder, and the active elements of electromechanical devices that perform a system of logic are shown on the rungs of the ladder. By following the rungs from left to right and from top to bottom, the sequence of operations can be documented. Additionally, if terminals are shown and marked, the actual interconnecting wires can be defined. Many servicemen use a ladder drawing to troubleshoot a circuit in preference to using an actual interconnection drawing because it helps them envision the flow of action through the circuit. The components that make up a logical circuit are switches, contacts, coils, circuits. These are included in the schematic drawing by using symbols to represent the components at the appropriate locations.

Notice in Figure 8-2 that some of the relay contacts are on the same line

while some are shown on parallel lines.

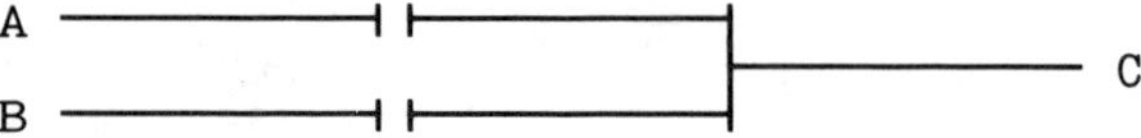

If either of the contacts on the same line is open, no current can flow from A to B; but if both contacts are closed, current can flow. This is called a series circuit. It is also called an AND circuit, because one contact AND the other contact must be closed to complete the circuit.

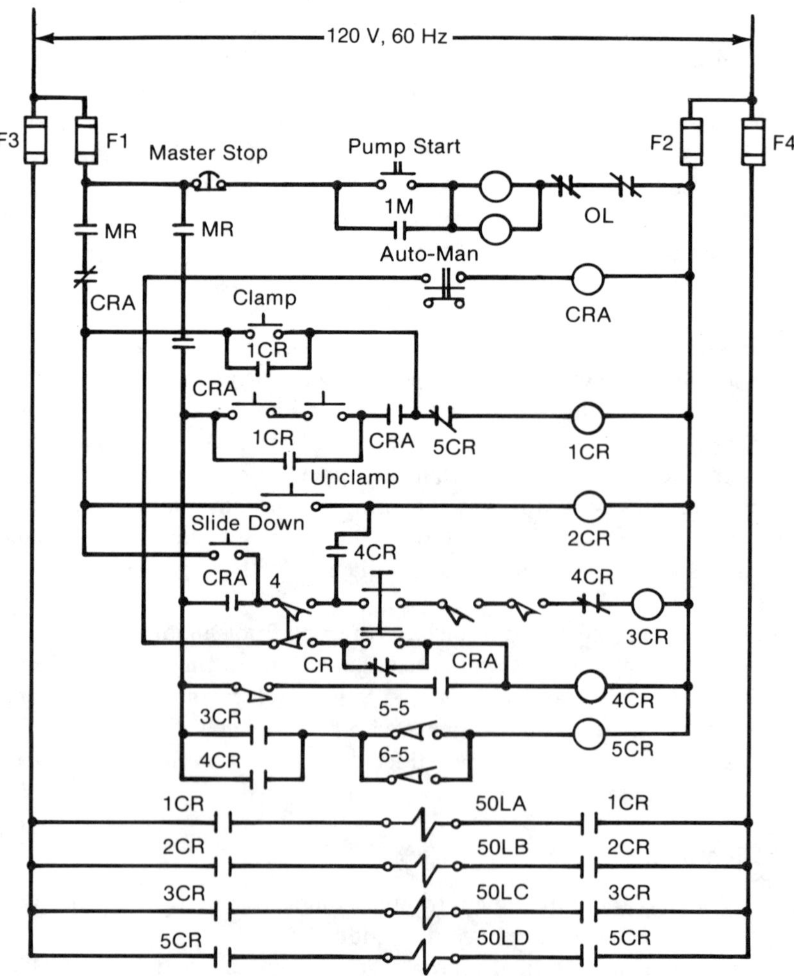

Figure 8-2. Schematic Wiring in Ladder Diagram Format

If either of the contacts shown on the parallel lines is closed, current can flow to C. Closing the upper one closes the circuit from A to C, while closing the lower one closes the circuit from B to C. This is called a parallel circuit. It is also called an OR circuit, because closing either the upper OR the lower contact can complete a circuit to C.

The contacts shown above are commonly designated "normally open". It is standard practice to show contacts as though the device containing them is de-energized. The other common form of contact is "normally closed", shown as ————————$\dashv$ / $\vdash$———————— , the condition it assumes when the device containing it is de-energized. In the energized condition, a normally closed contact opens, while a normally open contact closes. In order to read a ladder diagram one must be able to imagine the condition of the contacts as various loads are energized, because they will usually all be shown in the de-energized condition. The rule for draftsmen is, "Imagine the component in a box on the shelf, and draw the contacts the way they are in that condition."

Another common contact configuration combines a normally open and a normally closed contact:

This is sometimes called a normally open-normally closed contact, or a Form C contact. A Form A contact is normally open,

$$————————\dashv\ \vdash————————$$

while a Form B contact is normally closed.

$$————————\dashv\ /\ \vdash————————$$

When the device containing a Form C contact is energized, a circuit is closed between A and C, and the circuit between B and C is open. When it is de-energized, the circuit between B and C is closed, and the circuit between A and C is open.

8-5 Boolean Logic Diagrams

Logic can be shown in drawings by using Boolean logic symbols instead of ladder diagrams. Boolean algebra is a branch of mathematics that uses formulas to represent combinations of signal flow. It is possible by using the mathematics to design circuits that will satisfy logic conditions. Consequently, the symbols used in Boolean diagrams can be used to display the equivalent of ladder diagrams, and some designers prefer to use them. A few PLC's are programmed this way; the great majority, however, use ladder diagram symbology for programming.

The symbols commonly used for Boolean logic diagrams are shown in Unit 3 (Figure 3-4). Sometimes the symbols are called "gates". An AND gate, for example, performs the same circuit function as the series contact circuit shown above. An input is active and will cause the gate to function when there is a specific voltage applied to it. This may be zero volts, or it may be 1 volt or 5 volts, or –5 volts, but if that value is the one defined to activate the input, it will be active or "set" when the specific value is applied. Thus, in Figure 3-4(a), when A is set AND B is set, there will be an output from C. An inverse signal is indicated by the symbol

$$————————\!\circ$$

The OR symbol shows the function of the parallel contact path above. In Figure 3-4(b), when A is set OR B is set, there will be an output from C.

The Form C contact function described above could be diagrammed using the EXCLU-SIVE OR gate shown in Figure 3.4(d). This function allows an output from C when A is set, OR when B is set, but not when both A and B are set. Functions can be combined to illustrate complex circuits, just as the contacts in a ladder diagram do. When these combinations become complex, a knowledge of Boolean algebra is useful to find the most efficient arrangement of contacts.

Figure 8-3(a) shows a pump up/pump down circuit for a storage tank as it would be shown on a ladder diagram. When the START button is pushed, the circuit is activated. Relay CR is energized, and its normally open contact closes to maintain the circuit after the START button has been released. Pushing the STOP button will de-energize the CR relay.

When contact CR closes, a series circuit may be established through it and the normally open LO level switch, to energize relay 1CR. Its normally open contact 1CR closes to start the pump motor and fill the tank. As the level rises, the LO level switch will open, but relay 1CR will remain energized through the normally closed HI level switch and the 1CR contact. When the high level point is reached, the HI level switch contact will open, the pump will stop pumping, and it will remain off until the level has dropped again to the low level point.

Figure 8-3(b) shows the same circuit using AND and OR gates.

8-6 Screen Display

As programmed from a keyboard, and using a video screen to show the development of the program, the ladder diagram program display is usually a picture similar to that of the circuit desired, not a series of statements in a high level language. Programming done in accordance with Boolean expressions uses coded statements, not a diagram. Figure 8-4 shows the pump up/pump down circuit in ladder diagram form as it might appear on the programmer screen of one supplier's product. Other suppliers have similar but not identical displays.

Programmed using a Boolean-based code, the display might show the commands

```
IF 101 OR (K-1 AND NOT 102) THEN K1
IF K-1 AND (103 OR (K2 AND NOT 104)) THEN K2
M=K2
```

The inputs from the START push button, the STOP push button, and the level switches are connected to specific input points. These have numbers, and the numbers are the addresses of the registers where the input information (whether the switch is open or closed) is stored. The numbers may be used in the program to read or write information to specific registers. The programmer panel may have thumbwheel setters for defining numbers, or they may be defined by keystrokes. Each position on a line (ladder rung) is defined by giving it a number and pushing a key on the programmer that defines the function to be performed. Each key is marked with its function. Only a few keys are needed; the normally open contact in a series line, the normally open contact in a parallel line, the normally closed contact in a series line, the normally closed contact in a parallel line, a tee for branching circuits, the load (relay coil, solenoid valve, or motor coil); a counter, a timer, an adder, a subtractor, and a data transfer button are typically all that are required for the most complicated logic circuit. It is recommended that the information be organized by hand on a work sheet, which then becomes the reference for the information to be keyed into the memory of the processor, one line at a time, in sequence.

Devices other than contacts and coils, such as timers and adders, are programmed by moving to the position in the ladder diagram where the device occurs; pushing the button on the programmer identifying the device; and entering data that defines integers (seconds and tenths of seconds for timers, for example) and addresses (sources of inputs to add, or source of set and reset functions for timers, for example). CRT displays for an adder and a timer might show symbols similar to those of Figure 8-5.

This display illustrates a contact closure stored in register 201 turning on a timer stored in register 442. After 130 seconds the timer will develop an output voltage at output 7 that will energize a load. The timer will be reset when a normally closed contact sends a signal to input 313.

8-7 Other Functions

PLC's with functions other than relay logic are at some disadvantage if their only programming language is the ladder diagram or Boolean type. Although these are excellent for logic, they are difficult to use for servo positioning or for PID control. Consequently,

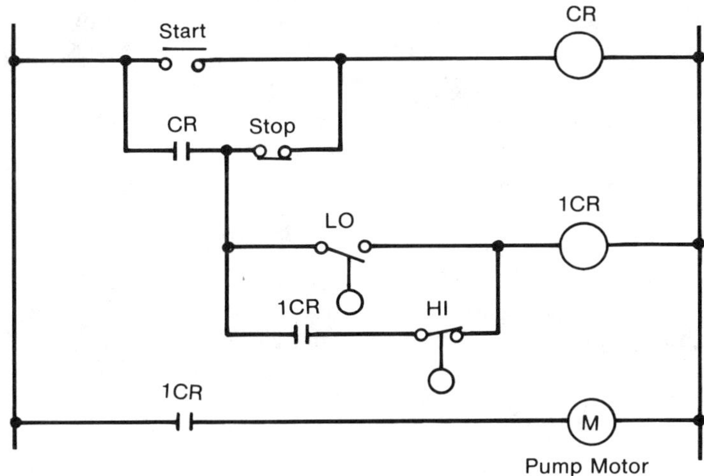

(a) Ladder Diagram

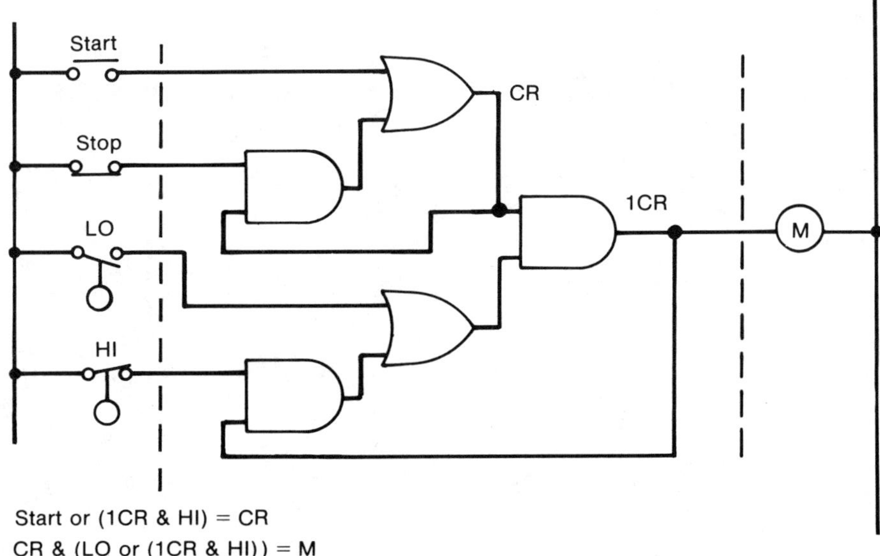

(b) Boolean Diagram

Figure 8-3. Pump Up/Pump Down Circuit

customized functional languages have been prepared that are specific to a supplier's product. Some equipment is offered with several languages, including ladder diagrams, mnemonics, variations of BASIC, languages specific to process control or motion control, and languages for creating graphic displays. High level languages, which can be worked into programming through an auxiliary computer, will be more flexible and less cumbersome. Today, however, the majority of PLC programming keyboards go only as far as a menu-selected display for PID configuration, with fill-in-the-blanks programming.

While the original and still principal application of programmable logic controllers is for discrete control operations, they can perform many other tasks that are optionally available in some of the larger models.

Drum Sequencers

Many operations require a combination of operations to be performed in steps. For the preparation of a batch of paint, for example, the process might consist of the following steps:

(a) Make sure pumps are turned on, valves are closed, and agitators are not turned on.

(b) Open valves A and B, add ingredient C for two minutes, and ingredient D until 200 pounds of it have been added.

(c) Turn on agitator, close valves, mix for 30 minutes.

(d) Open discharge valve, empty vessel.

(e) Turn on cleaning fluid, wash out tank, turn off agitator, shut off all motors.

All of these operations can be done with timers, counters, and discrete input and output scanning. A program that lists steps, actions to be taken, and parameters simplifies and coordinates the design and will usually save memory. Some sequencing programming permits adding jump and branch capabilities; that is, the capability to skip steps, depending on discrete input values or elapsed time, or the ability to perform one set of steps to satisfy one criteria or another set of steps to satisfy another criteria.

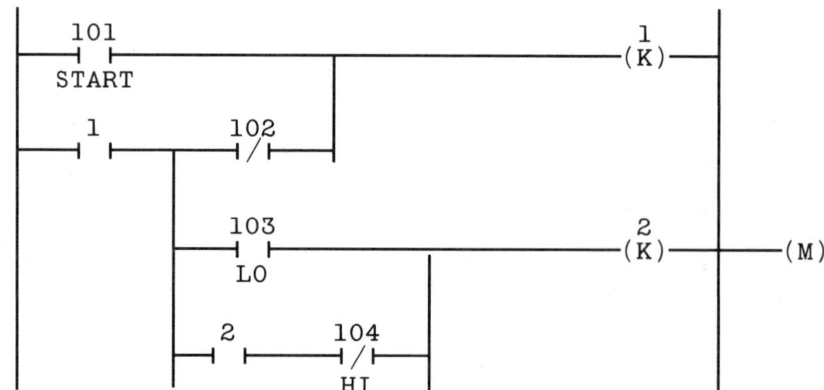

Figure 8-4. Ladder Diagram Screen Display — Pump Up/Pump Down Logic

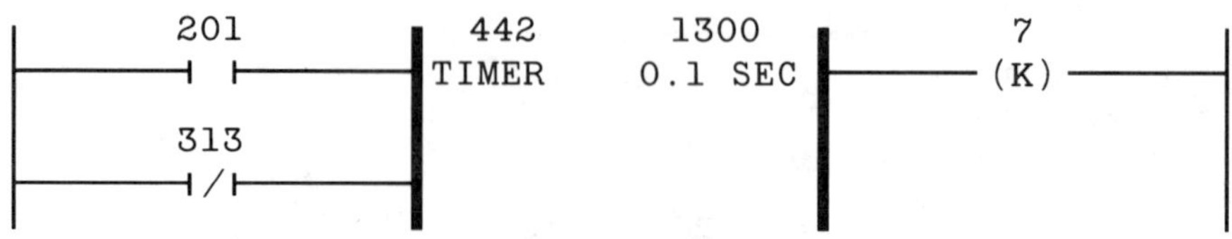

Figure 8-5. Representation of a Timer Function in Ladder Diagram Logic

Computation

Modules can be addressed that will perform the function of addition, subtraction, multiplication, division, and comparison. Some suppliers include special functions such as square root extraction.

PID Control

Using input modules that will accept analog signals and output modules that will develop 4 to 20 milliampere current outputs, programmable controllers are able to perform analog control operations. Internal programs can be written to scan inputs and use sampled data to perform proportional, integral, and derivative control based on the difference between measured process variable values and reference (set point) values. The details of this type of control (PID) are discussed in the preceding unit, in conjunction with the description of single-loop controllers, and the reader is referred for more detail to that section. By combinations of computing modules (summers and multipliers for gain adjustment, PID algorithms for cascade and ratio control, dividers for ratio adjustment), multiloop control functions can be realized and implemented. An example is given in Unit 12.

In addition to standard PID algorithms, there are extensions available from some suppliers that provide proportional servo control, point-to-point positioning, and integer/ratio master/slave positioning. These are used primarily in machine tool operations and approximate numerical control functions.

Data Transfer

Some suppliers have models that include the capability to store information in a matrix of registers, transfer this information into other matrices of registers, and search through the matrices, performing operations on the information in the registers. This is a very powerful capability, because it allows for data base management and information transfer capabilities that approximate the techniques of assembly language programming. The following activities may be available:

(a) Information can be stored in tables, and specific values "looked up". Specific values can be extracted and placed into registers for use in programs.

b) Data tables can be loaded, copied, and examined.

(c) Information in data tables can be searched by indexing through them. Values can be compared, biased, inverted, added, deleted, and used as pointers to other tables.

(d) System status information can be stored, and called up for examination.

(e) Matrix operations can be performed between two tables;

      For example, Logical AND of all matrix values

           Logical OR of all matrix values

           Logical exclusive OR of all matrix values

           Logical complement of one matrix

           Logical comparison of values in one matrix with those in another matrix

(f) Bit rotate operations permit masking and checking of matrix values serially.

8-8 Motion Control

Some suppliers have sufficient market for drives that control position, velocity, and torque of dc motors that they have developed units for applications using these parameters. Programs for these applications are most effective if they can be written in BASIC, or assembly language routines.

Many machining operations systems that use sequential control must also control the motion of rotating axles. For high precision products, closed-loop servos are being used. Control involves both positioning a rotating shaft and doing it at a controlled rate. Figure

8-6 shows the relationship between position and speed that must be satisfied to achieve motion control.

In one successful implementation, the loop starts with a position command. The actual position is measured by a digital encoder, or by a resolver, which is a type of rotary transformer. It develops an ac signal whose amplitude and phase are analogs of the shaft angle. The phase angle input is converted to digital form and compared to the position command, and an error signal is developed.

The PID output developed as a function of the error signal is cascaded into an inner loop controlling speed. This loop is used to compensate for load variables introduced by the motor and the connected load. For example, a hydraulic system has volume that influences response time and may have hydraulic resonance. An electric motor may have inertia and hysteresis.

The inner loop maintains velocity at a value that satisfies the position error. In many applications, this loop will be of the analog variety, because it must have high gain. The sampling time required for a digital loop could introduce disturbances. The velocity transmitter often used is a tachometer, directly coupled to the shaft to eliminate dead band. The tachometer creates a pulse string output, and if the loop is an analog one, this will be converted to a voltage.

8-9 Physical Characteristics of PLC's

PLC design permits operation in harsh environments to a greater degree than other types of digital control devices. This is indicated by the listing of typical operating specifications for PLC units as follows:

Operating temperatures — 0 to 60 degrees Centigrade
Storage temperature — –40 to 85 degrees Centigrade
Humidity — 5% to 95% RH, noncondensing
Vibration
      10-27 Hz at 1.3g
      33-100 Hz at 2.0g

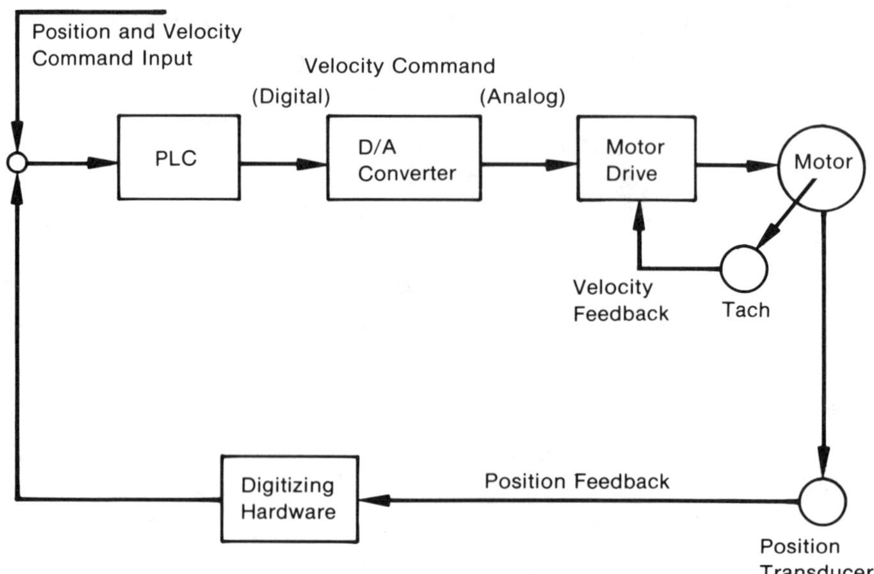

Figure 8-6. Motion Control Loops

Approximations — Size and Cost

I/O	Memory, Bytes	Cost Range, $
Small — 8–32	2K	500 to 5000
Medium — 32–512	4 to 7K	5000 to 10000
Large — 4096	up to 240K	15000 to 20000

8-10 Advances in Design

PLC's are a dynamic product. Every year new designs are offered, and options listed as new today will be superseded by the time this is read. Some of today's "new" features include the following:

(1) Intelligent I/O, in which the input/output modules include microprocessors that enable them to process the data they access and perform closed loop-control without the assistance of the processor.

(2) High speed color printout.

(3) Report generation modules.

(4) Participation in hierarchies of control.

(5) Mass storage peripherals.

(6) Backup, with millisecond transfer.

(7) Multi-tasking operation.

REFERENCES

1. Jones, C. T., and L. A. Bryan, *Programmable Controllers — Concepts and Applications*, International Programmable Controls, Inc. (1983).
2. Meinhold, Ted F., "Programmable Controllers . . . an Overview of Sizes and Capabilities Plus Guidelines on Selection, Installation, Maintenance, and Use," *Plant Engineering*, November 23, 1983, pp. 52–64.
3. Wylie, Mark E., "Intelligent I/O: New Smarts for PCs," *I and CS*, January, 1985, pp. 59–61.
4. Mellish, Michael T., "The Next Generation of PCs," *I and CS*, January, 1985, pp. 62–65.
5. Felder, Heinz F., and Gautam A. Tendulkar, "Multiprocessing Boosts PC Performance," *I and CS*, January, 1985, pp. 66–68.
6. Jasany, Lesie C., "Tying the Factory together with PCs and Networks," *Production Engineering*, April, 1984, pp. 52–63.
7. "Manufacturers Clear the Way for Integrating Control," *Production Engineering*, June, 1984, pp. 104–108.

EXERCISES

1. Propose a safeguard for diagnosing failure in a PLC I/O unit.

2. What are considerations in choosing a small PLC?

3. How can you determine memory and I/O quantity needed for a particular job?

4. PC suppliers prefer to sell hardware, not systems design. How can an inexperienced user put together a system of PLC's, networks, graphic terminals, computer links, etc.?

UNIT 9
Distributed Control

Distributed control refers to a system of digital instrumenta tion that is distributed geographically and also functionally. In the geographical context, a number of electronic assemblies are located in processing areas throughout a plant, wired to process-mounted sensors from which they receive information about the process conditions and also connected to process-regulating devices to which they send commands generated by programs that process the sensor-derived information. All the remotely located assemblies communicate over a communication network with a control room where operation is centralized. The operators working in this room use CRT displays to observe process and controller conditions. They use keyboards to send their commands to the system components, located locally and remotely, and to assume manual operation of process control loops if it becomes necessary to do so. There may be peripheral devices at the central location, including printers, links to computers, bulk memory, and other display devices. The communications network may correspond to any of the LAN's illustrated previously in Figure 5-3.

In the functional context, distributed control refers to the distribution of operating activities that the geographic distribution promotes. Each remotely located assembly is microprocessor-based, capable of collecting, processing, and distributing information individually. Failure of the centrally located equipment will not shut down the system. Local control operation will continue if the communication link is cut. Failure at a local installation will affect only one or, in the worst case, several loops of control; the data gathering and process control operations of the rest continue.

There are some process control system applications that only a minicomputer system can perform adequately and others for which single-loop control is the most practical choice. The characteristics of distributed control qualify its application for controlling process conditions to be somewhere between single-loop control systems and mainframe or minicomputer directed systems. As a preliminary to discussion of the position of distributed control in the list of devices that can perform closed-loop control, and in the context of advantages and limitations, the functions of the hardware at the local and remote locations will be reviewed.

9-1 Remote-Located Equipment

The equipment located close to the process areas receives inputs from process variable sensors and from discrete control devices such as limit switches and relay contacts. Inputs from discrete control devices are ac or dc voltages transmitted over twisted pairs of wire. Inputs from process variable sensors at the present time (1985) are analog, 4 to 20 milliampere or 10 to 50 milliampere signals transmitted over shielded, twisted pairs of wires to terminal board connections. Transmitters that produce binary coded outputs have not yet been used, although it is probable that they soon will be.

The input and output signals may terminate in a cabinet dedicated to that purpose but are more often taken directly to a housing in which are mounted terminal boards, power supplies, and the electronic assemblies that perform the digital control, information transfer, alarm detection and generation, and other computing functions of the device. This housing will usually be gasketed to keep out moisture, may be supplied with fans for removing heat generated by the power supplies, and will have a NEMA 4 or NEMA 12 type of construction. It should be located in a protected area that observes the limits of temperature, humidity, vibration, and corrosion defined for the electronic equipment it contains.

The electronic assemblies will be mounted in some sort of rack formation, holding printed circuit cards to perform the functions of processing inputs, transferring information, maintaining files of data about individual functions and groups of functions, performing computations and diagnostics, and so on. A number of microprocessors divide the tasks. Interconnection in the cabinet between electronic assemblies is often done with multiconductor cabling that terminates in plug-type connectors.

Each electronic assembly includes time-shared circuitry that will accommodate a number of input signals and a number of output signals. This means that one assembly may be assigned to provide more than one control loop. Some suppliers endorse the concept of single-loop integrity and will provide control from a single module for only one output, assuring that if there is an electronic failure only one process control loop will be affected. Other suppliers provide control for as many as 30 loops from one assembly, although eight or sixteen is a more common number.

The programs that define the operations at the remote locations are stored in memory on the circuit boards that make up each assembly. Programming that customizes each assembly to the particular process operations that it has been selected to control is done from the central control room over the communications network, often by a fill-in-the-blanks manipulation from the keyboard. This program is commonly stored on disk in the central operating station electronics package. The word "configuration" is often used for this customizing of the system operation. This term will be used in this section to differentiate it from the programming that makes the system perform.

The programming of the remote-located unit allows it to function as a data management system. Functions are time-shared by moving information from registers into a RAM working area where computations are performed. Memory storage areas are assigned to hold information about input values, output values, the subroutines used for computing control values, and the parameters that are specific to each control loop. Analog inputs will be scanned at regular intervals, conditioned, converted to digital values, and the binary equivalents updated into registers. A typical number of inputs is 30, and a typical scan time is ½ second. During this same time interval, all the output values will be computed on the basis of current information and stored in another set of registers. The scan time period is divided into a number of smaller periods, of lengths ranging from a few milliseconds to perhaps 30 milliseconds. During each of these "time slots" a program with the following elements is carried out:

(1) Move into the working area of RAM the subroutine to be performed. A subroutine might be a portion of the program for a PID controller, the error detection comparison between set point and measured variable, for example.

(2) Move into the working area the parameters that have been configured and saved to define this particular subroutine; gains, limits, ranges, units of measure, and so on.

(3) Move into the working area the latest update of input value information for the inputs that are specific to the loop to which the time slot has been assigned.

(4) Perform all the computations called for in the subroutine, to develop updated output values.

(5) Store the new outputs in registers and continue with the next time slot.

This sounds like a lot of action for a fraction of a second, but a command requires only one or two millionths of a second to perform, so it is an entirely practical concept.

In addition to the computational programs, which are the principal activity of a controller file, background programs are continuously running diagnostic routines, bulk memory is retained for transferral of data values for trending, and alarm signals are generated based on configured process variable value limits. When there is a requirement for a large amount of data acquisition, it is common to use a separate data logger for collecting this information and routing it into the system through a serial port in a controller file.

9-2 Central Control Area Instrumentation

The remote controller files communicate with the equipment that makes up the operator's station in the central control area. In its simplest form, this will consist of a keyboard used to address the system for entering commands; and a CRT monitor on which the information about the process conditions is displayed. These may be built into an elaborate console of the type illustrated in Figure 9-1. Electronics hardware and software for the entire system may be built into the console, or housed in separate cabinets.

The CRT interface is important to productivity, because the operator may not be near the process area. He must depend on the information displayed on the CRT screen to tell him what is actually going on. The variety of display types available to the operator is the feature of distributed control that makes it different from all other forms of digital control equipment.

At least six types of CRT display provide information for the operator. Three of these, the detail display, the group display, and the overview display, are related to the time slots previously discussed. The other three are the graphics display, the trending display, and displays of lists. The display of most interest to the operator will be the group display, which shows the operating conditions for a group of related loops. Figure 9-2 illustrates a portion of a group display.

The individual displays in the group simulate the appearance of single-loop controllers. The operator, if he has been accustomed to panel-mounted analog instruments, will be familiar with the visual layout of the picture and also with the uses that he can make of it. Looking at the analog instrument he can see dials or bar graph indicators that show the value of the process variable, the percentage of output signal, the set point value, and whether the instrument is in manual or automatic mode. The CRT display shows the same information in a grouping that is very much like that of the analog instrument (see Figure 7-7, a typical single-loop controller). A bar graph increases its height as the process variable value increases, and the same display arrangement appears for the output signal percentage. Below the analog controller there will usually be a plastic nameplate on which will be engraved a tag number identifying the service of the instrument; this same information appears on the CRT display.

The panel-mounted analog controller will have a switch mounted on the front panel for selection of the manual or the automatic mode of operation. It will also have knobs or push

Figure 9-1. Distributed Control System Console

(Courtesy Beckman Instruments, Inc.)

buttons so that the operator can change the output value when the manual mode has been selected or the set point value in either mode. For the distributed control system, which has only a picture of a controller, this same mode selection can be made from the keyboard, which communicates with the CRT through a cable connector. Using the keyboard, the operator can select a specific time slot display, change its operating mode to manual or automatic, and enter the digital value of the desired output or set point. Newer systems are using CRT screens that implement the changes when the display is touched in the proper place and sequence. "User friendly" is a somewhat overworked term used to indicate that digital equipment is easy to work with. A lot of thought has gone into distributed control displays to make them "operator friendly".

Limits can be configured into the time slot parameters to generate alarms if deviation from set point, a limiting value of output signal, or a limiting value of the process variable is exceeded. The ability to select these alarm points is a part of almost every supplier's configuring capability. Usually, if one of the limits is exceeded, the alarm condition will be indicated in a way that will attract the operator's attention by a blinking display or a background color change. An audible signal may be included also. This duplicates the action with which the operator will be familiar from his analog control instrument panel experience, where an annunciator window will light up and blink to indicate an alarm condition. There may also be an audible alarm to alert him to the condition. The operator's instinctive action to an alarm signal is to push an acknowledgment button to silence the audible signal. To the same end, there will be an acknowledgment button on the keyboard, and when the operator pushes this, the display will change from the group display to a detail display showing the parameters for the specific time slot in alarm. Figure 9-3 illustrates a typical detail display.

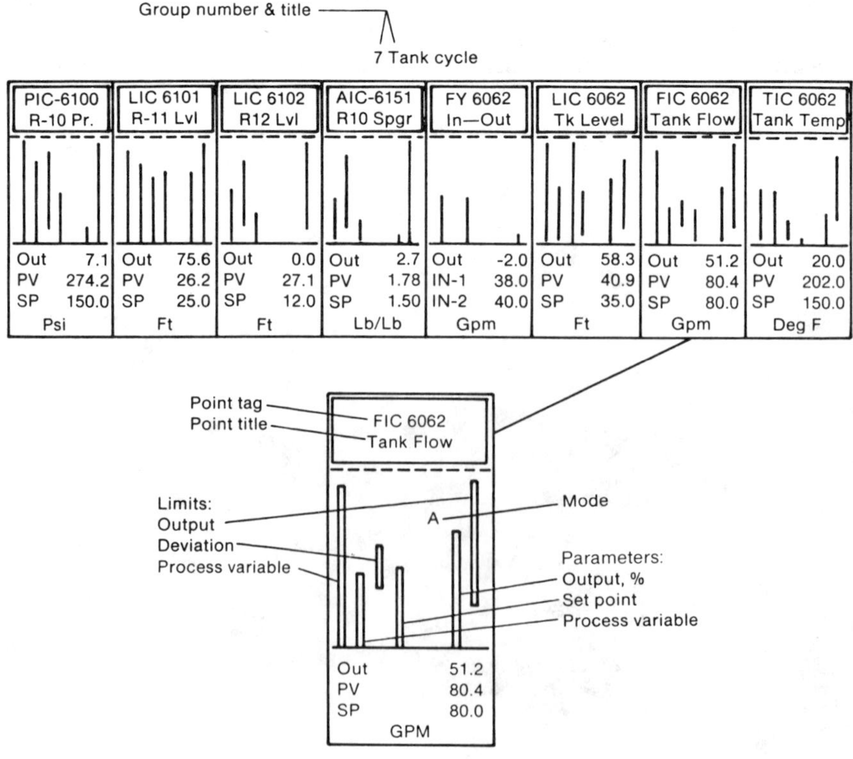

Figure 9-2. Group Display

(From *Understanding Distributed Process Control*,
J. A. Moore and S. M. Herb, ©1983, Instrument Society of America)

The most basic of the displays, the detail display is the means for defining the program for the time slot. The symbols and numbers shown in the illustration represent the parameters of the time slot and are entered from the keyboard as part of the configuration of the time slot. They define the type of function the slot is to perform (the CONF word in Figure 9-3); the source of the input values (the CINP words); the operating parameters, which individualize the function (K values for tuning constants and range); and the operating limits and rates. These values are entered from a keyboard when the system is being generated; it may be the same keyboard used by the operator, or it may be a special keyboard used only by the design engineer for programming. Under any circumstances, there will be a key switch on the keyboard that must be unlocked for this type of programming to be done. Normally the operator will not have access to the key; his use of the keyboard will be restricted to set point change and output change (in manual mode). Additional security measures may prevent the operator from taking even these actions.

The third display always used is the overview display, illustrated in Figure 9-4. This display shows the outlines of a number of groups (twelve, in the illustration). Only the indication of deviation from set point for each of the time slots included in each group appears in the overview display. This is intended to give the operator a birds-eye view of his entire area of responsibility. The intent is to pinpoint abnormal operation, because if every detail in a group is operating at set point all that will be visible will be a line across the picture of the group; but if there is a deviation that exceeds a configured limit above or below set point, it will show as a vertical line extending above or below the horizontal "normal condition" line and will stand out, showing the operator where potential trouble spots are. In many designs there is a hierarchy of alarm indication that extends to the other displays. An

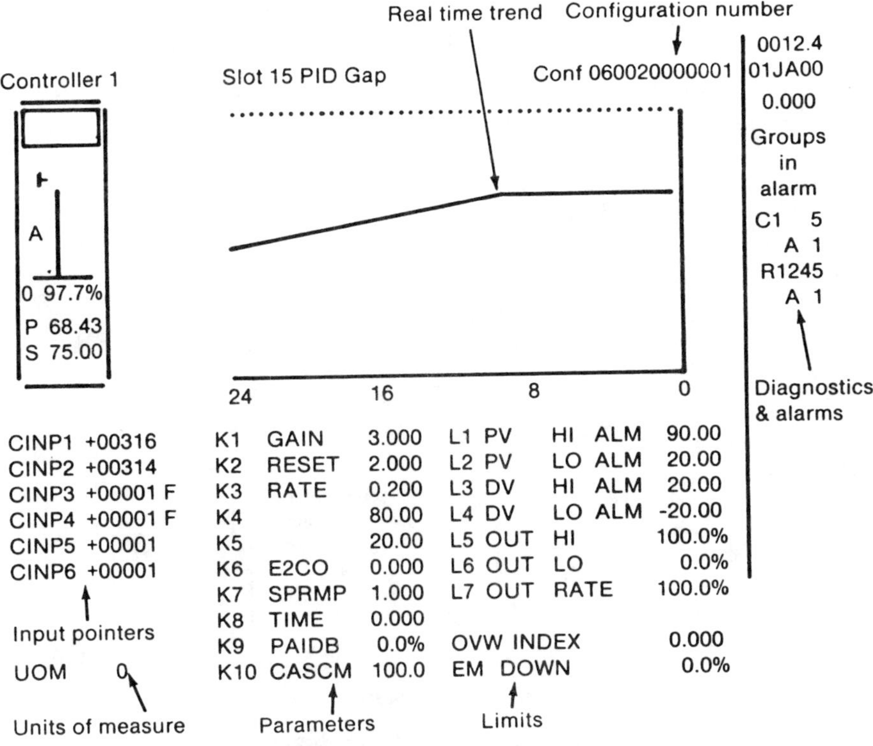

Figure 9-3. Detail Display

(From *Understanding Distributed Process Control*,
J. A. Moore and S. M. Herb, ©1983, Instrument Society of America)

alarm condition will be indicated by a change in color when the vertical "deviation out of spec" line reaches a limit. When the alarm is acknowledged, the display will change to the appropriate group display. A second acknowledgment will call up the detail display of the slot where the alarm occurred, so that if the operator needs to check limit settings or find out more information about the programming of the time slot, he does not have to hunt for it or spend time calling it up on the screen.

Of the three displays described, the group display is probably the most useful to the operator because it presents a capsule view of everything going on in a portion of the process.

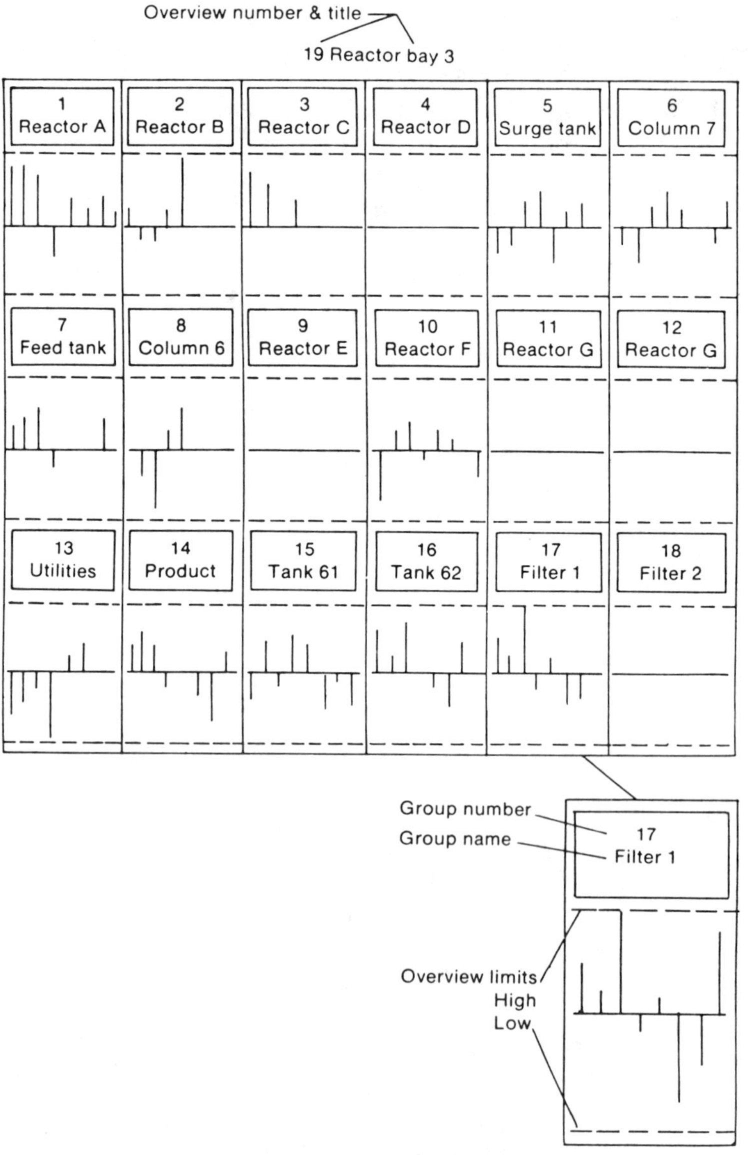

Figure 9-4. Overview Display

(From *Understanding Distributed Process Control*,
J. A. Moore and S. M. Herb, © 1983, Instrument Society of America)

The key to its importance is that a group display is created by configuration, and any combination of functions within the limits of the number of slots that can be displayed can be put together in any relationship desired. This flexibility is one of the outstanding qualities of distributed control: from the keyboard, parameters and configurations can be changed easily and quickly. As long as there is an input value connected and available for scanning, changes can be made in control strategy to try out ideas, mistakes can be corrected, and displays added or removed without touching a tool or adding pieces of hardware.

Graphics displays, trend displays, and lists may be created on the CRT screen. Some lists are always included in the basic programming; others have to be configured. Figure 9-5 shows a listing of input values. Some variation of this list will appear in all systems. The list provides a place for programming the range of the signal input as an analog voltage, a place for defining high and low limits, an identifying configuration word that defines the analog

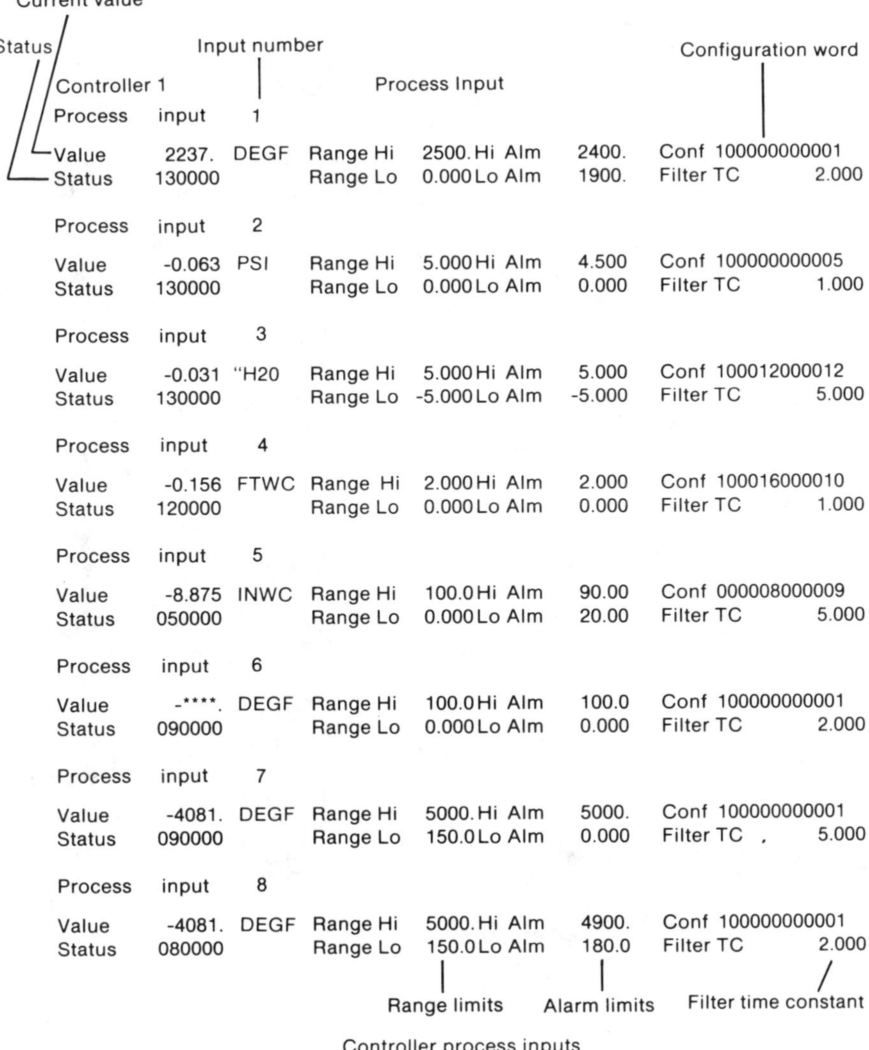

Figure 9-5. List

(From *Understanding Distributed Process Control*,
J. A. Moore and S. M. Herb, ©1983, Instrument Society of America)

value (1 to 5 volts, 0 to 4 volts, etc.), and a level of filtering that can be applied to noisy signals. Other standard lists might include dictionaries of messages, units of measure, menus, alarm summaries, status reports of highway conditions, and data base contents. Created lists could include messages to indicate corrective action to an operator in the event of an alarm situation, and indexes of created displays for operator reference.

The graphics display is optional in most systems but it should be included in the system equipment package if the economics of the job can support it. A graphics display, in its most useful and imaginative form, is a special kind of group display in which the time slots assigned to a specific portion of the process are shown on the screen in relation to a picture of the process. A typical graphics display is shown in Figure 9-6.

Only a limited number of loops of control can be shown in one graphic display without creating a confusing picture, but with care and imagination it is possible to give the operator a picture that will help him in his work. Displays can be dynamic. Tanks can be made to empty and fill, valves can be made to change color to show open and closed positions, and the observing operator can see changing relationships between variables as they occur. In some graphics displays, the time slot symbols can be addressed directly, just as in a standard group display, so that set points and output values can be changed directly from the picture on the screen.

An example of another helpful display is shown in Figure 9-7. This displays trends of the values of several variables over a period of time. Suppliers vary in their design of the trend display but all have it in one form or another, with a variety of time scales that can be as small as 90 seconds or as long as seven days.

In most distributed control systems, trended information can be archived, collected on hard or floppy disks over a period of time, for review at a later time. Several systems cause

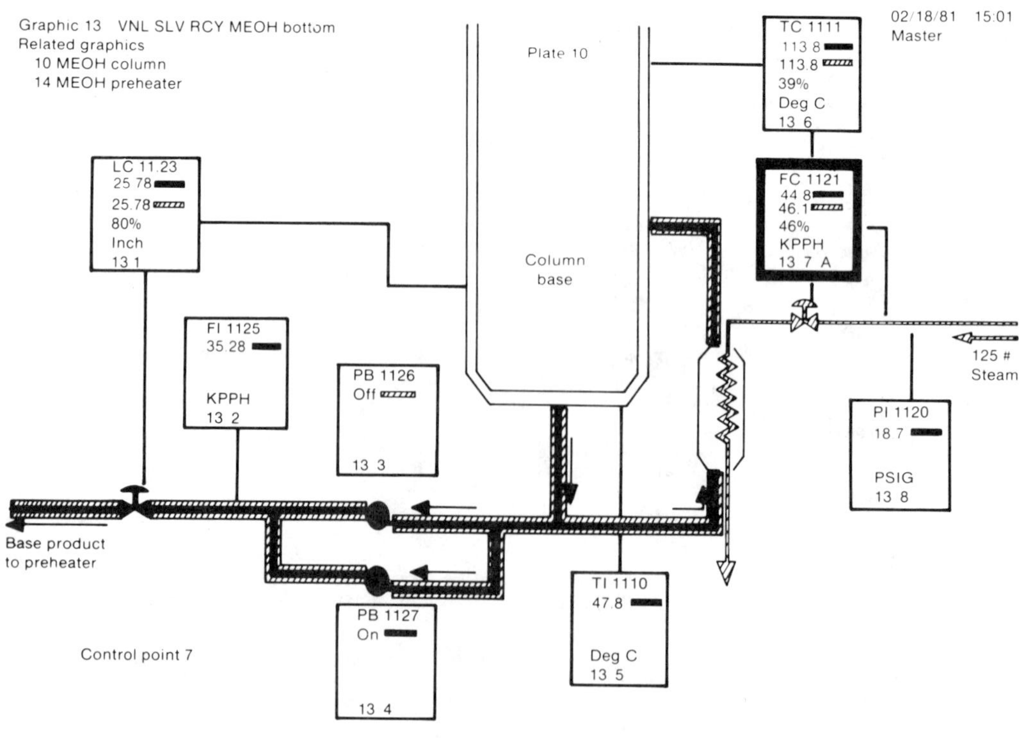

Figure 9-6. Graphics Display

(From *Understanding Distributed Process Control*,
J. A. Moore and S. M. Herb, ©1983, Instrument Society of America)

archiving to begin when there is an alarm condition, including records for a period before and after the upset. This record is valuable when a number of events can happen interactively and almost simultaneously. In such a case, it is important to know which was cause and which was effect.

9-3 Configuration

Just as configuration is an alternate term for programming often applied for distributed control systems, the word algorithm is encountered frequently, applied to the subroutines that are used for control functions. Algorithms are subroutines that can be assigned to time slots, linked, and used for control or for computation. Some suppliers of distributed control equipment break the functions down into basic operations. Figure 9-8 shows a PID control operation constructed (in software, by configuration) by a combination of blocks of subroutines that represent the input signal storage and conditioning for a process variable; the establishment of alarms from a high limit and a low limit to the input value; the generation of a reference signal, or set point; the comparison of the two to develop an error signal; the development of an output signal that is a function of the amount of deviation, the length of time the deviation has existed and the rate at which it is changing; the generation of a manually developed output; a switch for selecting the manual or automatic output; a block for making the same selection from a digital signal; an output block for the digital signal; a block for the output analog signal storage and conditioning; and a limiting function to prevent the output signal going below 10%. Each of these blocks has a name, and configuration gives them an address. Configuration points the input of a block to the output of another block to link them together. Many varieties of PID control can be made in this way.

The alternative is a more complicated time slot, illustrated in Figure 9-9, which has already done the linking in software to create a more complete PID algorithm. One time slot

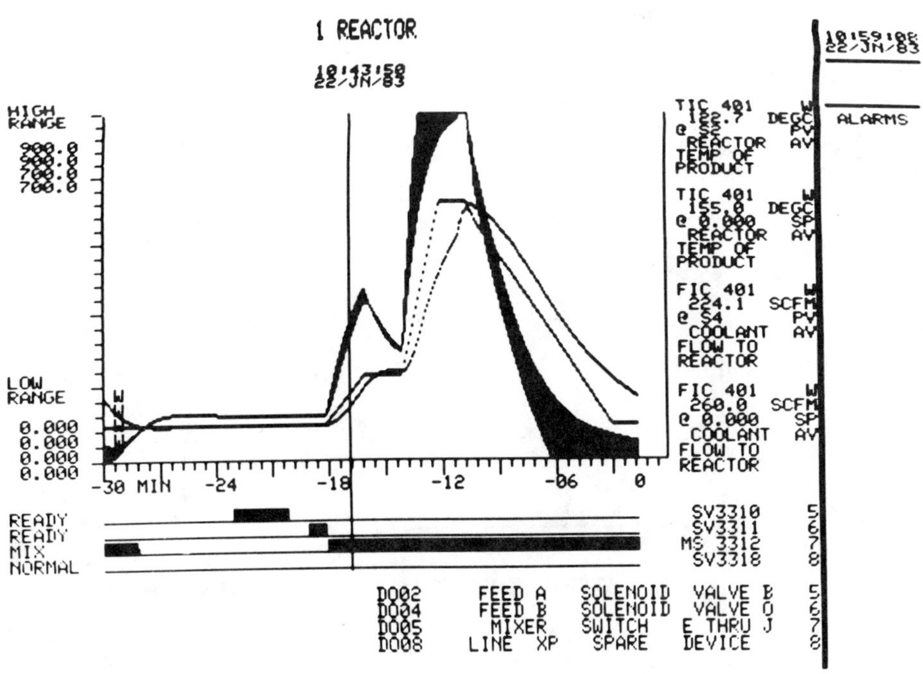

Figure 9-7. Trend Display

(From *Understanding Distributed Process Control*,
J. A. Moore and S. M. Herb, ©1983, Instrument Society of America)

is used to perform the functions done in eight in the preceding example. Here the set point is internal, and so is the manual operation. Process variable and digital inputs already processed and in storage are addressed. Linking with other algorithms can be done by configuration to create larger pieces in several time slots. A cascade control operation, for example, can be put together in two time slots by pointing the address for the source of the secondary PID set point to the output register of the primary PID algorithm. In some systems, a cascade control system can be configured in a single time slot.

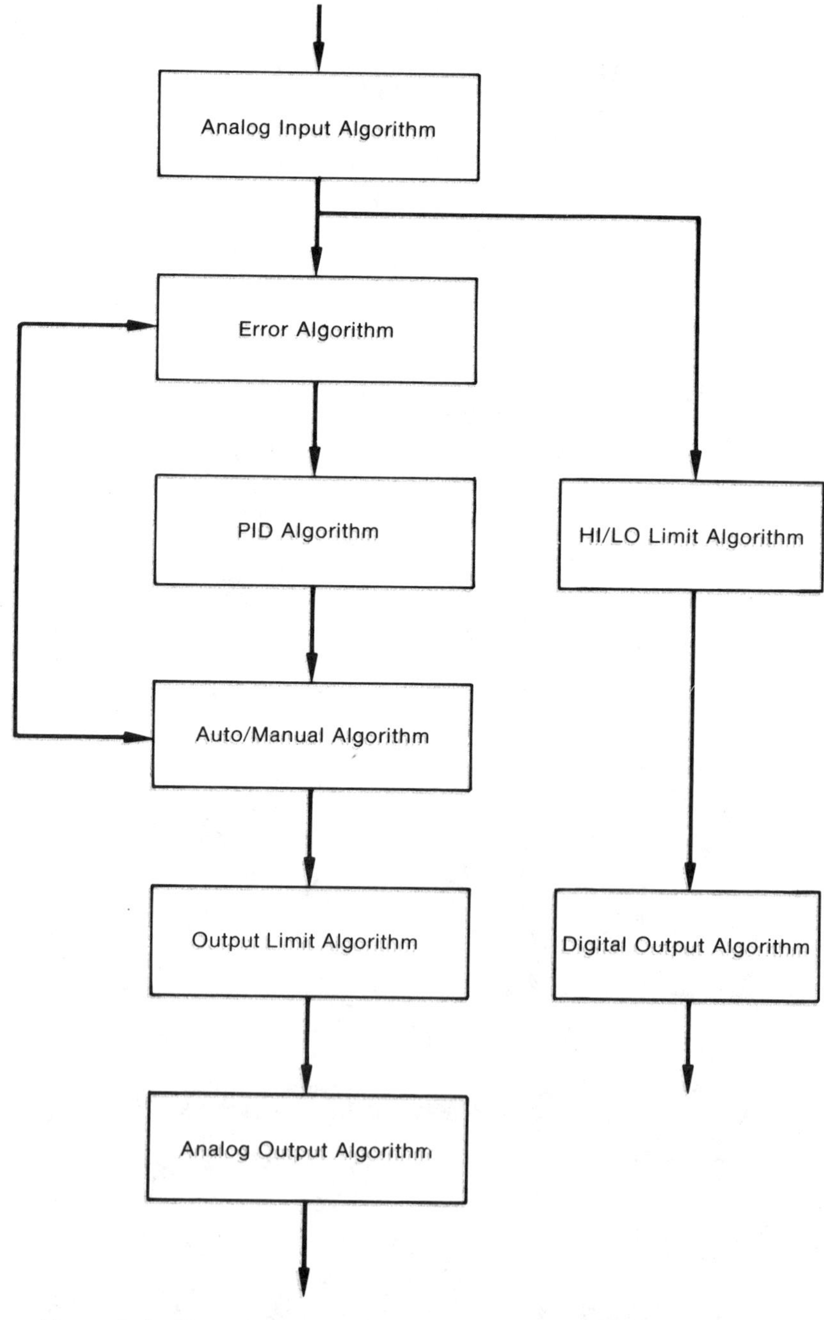

Figure 9-8. Algorithms Corresponding to Individual Subroutines

Proponents of the two types of configuring tend to prefer the type they are used to. Configuration is simpler using the more complete algorithms, but having a larger number of more elementary ones increases the flexibilty of application and can result in more imaginative configuration.

Algorithms provide many functions. A library of algorithms, including those used by most distributed control suppliers, is listed in Table 9-1.

Table 9-1
Commonly Used Algorithms

PID Algorithms	**PID Functions**
PID	Error Generation
PID — Ratio	Output Limiting
PID — Floating	Output Tracking
Auto/Manual	Feedforward
Supervisory Control	Anti-Reset Windup
Adaptive Tuning	Set Point Clamping
	Set Point Ramping
Computation	**Input/Output**
Summer	Analog Input
Multiplier	Analog Output
Function Generator	Digital Input
Dead Time	Digital Output
Median	
Lead/Lag	**Logical Functions**
High Select, Low Select	AND, OR, EXCLUSIVE OR
Mass Flow	Timer
Square Root	Counter
Log Function	Flip-Flop
Exponential Function	Sequencer

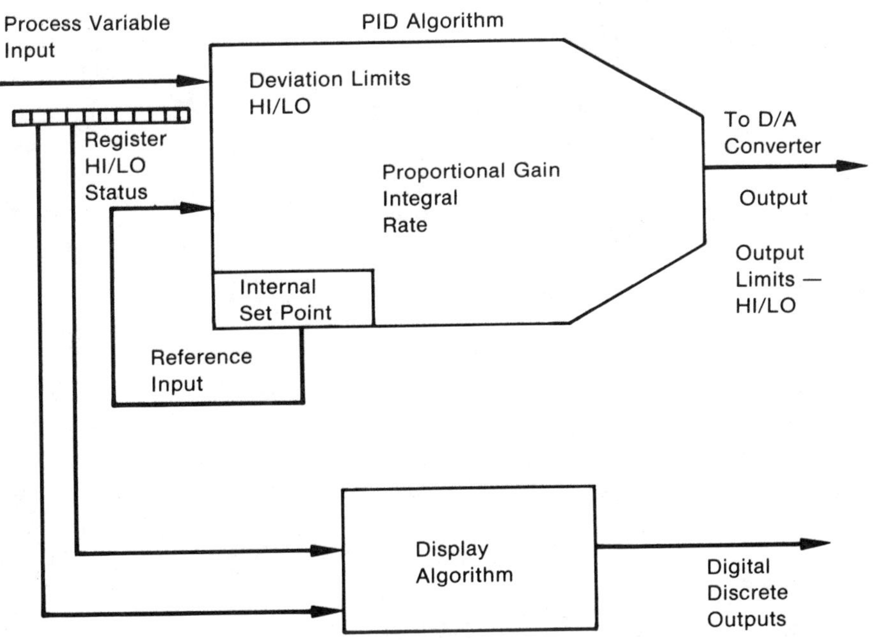

Figure 9-9. Algorithms Combining Subroutines

Having this choice of function means that the control system designer can build his controls to suit his production process response and capacity requirements much more easily than if he is confined to using discrete analog hardware. The desirability of computing process variable values that cannot be measured directly has been referred to previously. Using computing algorithms, this can easily be done by configuration in a distributed control system. A computation of energy, for example, can be configured from flow and temperature signal values and used as a process variable, as shown in Figure 9-10.

The other benefit of the flexibility supplied by configuration is that pointed out in the previous chapter as an asset of programmable logic controllers. Mistakes can be easily corrected, and changes can be installed and tried without any hardware being involved. Something is always wrong or is found to have been overlooked when a system is started up; the ability to correct the mistake from the keyboard saves much time, money, and embarassment.

9-4 Advantages of Distributed Control

Distributed control is still a new control form; the first commercial installations were made less than ten years ago, and today there are almost 100 suppliers of this type of equipment. The delineations of hardware and displays given above are background for an explanation of the reasons for the success of distributed control systems as a replacement for analog equipment to perform process control. There are several distinct reasons. They include benefits for the person buying the equipment, benefits for the operator, and benefits for the designer of the control system.

It must be pointed out that some of these benefits are provided also by programmable logic controllers when they are connected by a data highway to a central operating area.

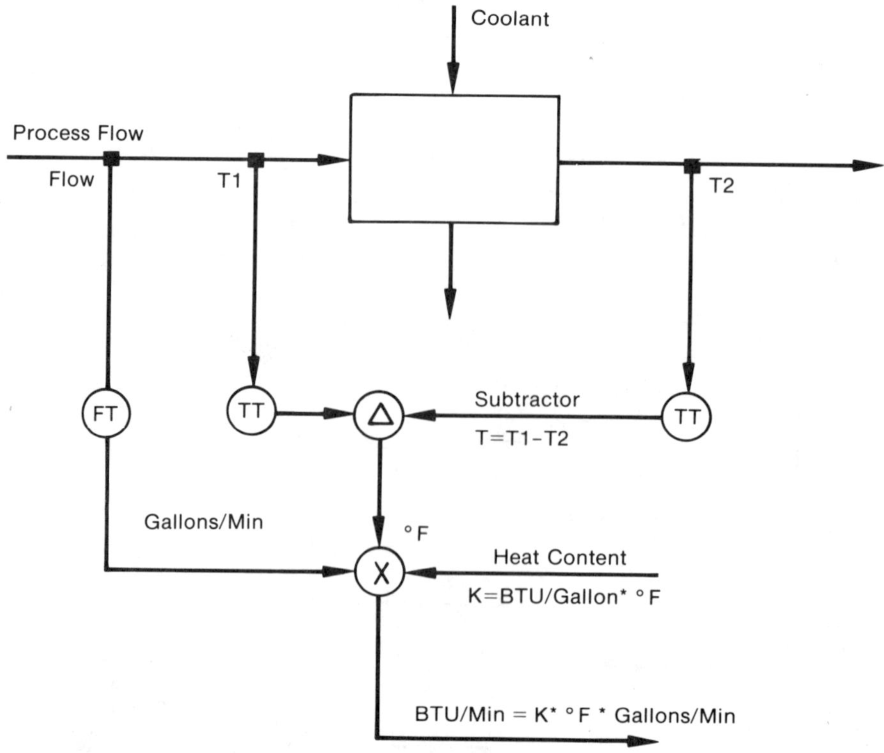

Figure 9-10. Btu Computation

The benefits for management, the source of decisions to spend money for new equipment, is in savings and in improved production. The immediate consideration is always in savings, and it can be demonstrated that installation costs when distributed control is used can be lower than for an equivalent system of analog instruments if the size of the system purchased is large enough. The break point is between 20 and 30 loops of control. Savings are derived from reductions in engineering costs, in wiring costs, and in panel costs.

Wiring Costs

For an analog system, a central control room is necessary, or at least a large panel must be used to mount controllers, indicators, recorders, switches for motor control, and other analog and digital equipment. Every device must be connected by wiring to the field-mounted sensor, regulator, relay, or other piece of equipment to which it is related. These wires will have to be run in any case, but it is much less expensive to run them to an electronic assembly mounted locally, close to the processing area, than it is to run the wires all the way back to the control room. The communication between the locally mounted electronic assemblies and the operator station in the control room will be over a coaxial cable or fiber optic cable, which may be two or more miles in length. One data highway may take the place of many hundred metallic wires.

Panel Costs

It cannot be argued that the operator station CRT displays will replace an instrument panel completely, because almost any installation will have the need for mounting some recorders, switches, annunciators, etc. However, the CRT displays can provide the same information as many of the panel-mounted types, and the size of the panel (and the engineering time required to design it) can be reduced. A significant saving occurs if signal conditioning is done by software at the remote locations. For analog system signal conditioning, back of panel devices must be mounted and interconnected with wiring. This is costly in panel space, engineering design, and wiring connections.

Other benefits to management from improved production derive from the benefits to the operator and to the process designer.

Accessibility of Information

Anyone who has operated from an instrument panel can confirm the validity of the following example of operator practice. The number of instruments on an analog instrument panel that the average operator has to be aware of is often too great for optimum use. Imagine an operator keeping track of twenty to thirty control loops, with indications scattered over ten or fifteen feet of panel space. When everything is behaving smoothly he will operate by keeping an eye on the indicators, not really watching any of them but knowing from experience that all measurements are within acceptable limits. Only when there is a change to be made or an emergency to be brought under control does he have to control manually, change set points, and watch closely the interactions of a number of loops. The problem is that when he stands close enough to the panel to read the indications, he can only see one or two devices clearly; others equally important may not be in his angle of vision. When he stands back far enough to get a picture of everything going on, he is not close enough to make changes. This is the condition that the CRT displays described above have been designed to improve, and this is why the composition of a group display is so important. By being able to sit at a screen that shows a group of operating conditions, organized to suit his needs, within reach of a keyboard from which he can enter commands for mode change and corrective action, the operator's usefulness increases. The operator benefits from distributed control because it compresses his working area from a wide spread of panel-mounted instruments to the concentrated convenience of an airline pilot's cockpit.

Flexibility

Benefits to control system design and from that to improved operation are obtained in part from the flexibility provided by configuration. There are two aspects of this flexibility. One is that configuration promotes the ability to change design to fit process conditions very easily. The other aspect is that supplier standards have provided a rich array of control functions for the designer to select from for the solution of simple and difficult control problems.

9-5   Reliability

The dependability of distributed control also affects the choice between it, single-loop control, programmable logic control, and control using a minicomputer. Reliability of any digital device is a function of time between failures, and time to rectify a failure. Burn-in procedures and quality control by most suppliers have resulted in time between failure periods of approximately 100 days for the CRT-based operator station section of a distributed control system and perhaps three times as long for the remote controller portion. Better numbers can be assured by using redundant equipment, external and internal backup, and backup power supply equipment.

Some redundant circuits and hardware may be included in the standard design of most suppliers' equipment. The amounts vary, and it is more common to see redundancy available as an option. Most suppliers include a redundant data highway as a standard design feature, because its integrity affects the entire communications network. Some suppliers include redundant power supplies and processing circuits, but these are more commonly optional. Redundancy and backup are not free. How much it is advisable to have is a function of the cost of failure. Failure of a control system for a reheat furnace, for example, is not catastrophic. The steel is recoverable and, indeed, will hold its heat for a long time, probably long enough for repairs to be made to the control system. On the other hand, failure of a control system for a batch reactor making plastic material that will harden in the piping if it stops flowing could result in the loss of an expensive piece of process equipment. There is almost no time allowable for repairs in such a situation.

The first line of defense against failure is the background program for diagnostics that is a part of most systems. When tests (time out of dead man timers, failure of check sums to produce the correct values) signal a possible fault, the condition will be indicated on the CRT screen, usually in a coded message. The message may give a clue to the location of the problem, but it is important to have a supplier or a service man who is familiar with the equipment immediately available. The usual recommendation is to keep a supply of spare, reliable printed circuit cards on hand, replace cards until the system can be reset, and resume operation. An excellent rule, then, is to have spare parts on hand and someone who knows what to do with them.

Another means of increasing reliability is backup. There are a number of ways to back up distributed control equipment:

(1) Purchase duplicate electronic equipment that will automatically and transparently replace defective controller files, or a part of them, if there is a failure.

(2) Provide a separate device that can supply a manually adjusted signal to open or close the final regulating device to its failsafe position, the position that it should go to if all power fails.

(3) Route the output signal through a separate analog controller that can be used to maintain control on its own merits if the distributed control system output fails.

A user seriously concerned with the results of failure will take steps to maintain the source of electrical power. Like an automobile, distributed control equipment will not function if the dc power supply fails. Redundant, auctioneered (the one with the higher output is dominant; if one fails, the other automatically takes its place) dc supplies are usually optional. An entire system supply can be backed up by batteries, charged continuously while the alternating power source is functional, and switched in automatically to maintain power *for a time* if the ac fails. Again, the amount of time during which the system must be reliable is a function of the amount of backup time purchased. This type of "uninterruptible" power supply is probably the most dependable way to assure reliability.

9-6 Bases for Selection of Control Equipment

Distributed control systems have many benefits, but they have limitations also. Chief among these are the following:

(1) The process variable value displayed (the value from which the process is being controlled) is always a little bit behind the real-time value of the process variable. There are a couple of reasons for this. One is that there is a scan time, and during most of it a variable is in a hold state, waiting to be serviced. This is a concern only for very fast processes. Secondly, the updating of displays is affected by the number of stations serviced by a highway. As more and more devices are added and the amount of information transfer multiplies, delays in updating screen display may become apparent. High baud rates and good design can help alleviate the situation. Signals, for example, may be transmitted on an exception basis and not be updated unless they have changed significantly from a previously read value.

(2) Just as PLC's do a good job of displaying logic on ladder diagram-like displays but have difficulty in using the same display for analog functions, distributed control displays satisfy operator needs for PID control but do not exemplify logical design very well. Designers of displays have tried to convey logical status by using words and symbols, and an operator can become familiar with the appearance of proper operation. A lot of imagination is required to figure out just what is happening in a logical circuit that has many branches. It must be admitted that in the hard-wired implementation of relay circuits, where the status light is the only indication of an on or off situation (except for pinball and slot machines), there is a similar groping for information.

(3) The tendency of algorithms to be control and computation oriented rather than process oriented is a constraint based on the volume of sales for any specific application. A companion limitation is that there is no easy way for a user to create new algorithms.

What, then, can be said about the relative place for distributed control when choosing between single-loop controllers, programmable logic controllers, distributed controllers, and systems based on minicomputer or mainframe digital computers. The following are general rules. It must be remembered that all rules must be qualified for the circumstances of any specific situation.

1. If the system is for PID control and there are no more than thirty loops, use single-loop control.

2. If the majority of the control requirement involves digital logic with a small amount of PID control, use a programmable logic controller.

3. If the requirement is for a large amount of PID control, particularly if a number of the loops interact, consider distributed control.

4. If process oriented algorithms or process modeling are required to get best results, or if optimizing is a requirement, consider distributed control with a computer as part of the network. However, if the process is one where special-purpose programs for minicomputer and mainframe computer systems, as well as specialized sensors, are commercially available, review the performance of installed systems that are using them and consider them carefully for your own application.

REFERENCE

1. Moore, J. A. and Sam Herb, *Understanding Distributed Process Control*, Research Triangle Park, North Carolina: Instrument Society of America (1983).

EXERCISES

1. At least two CRT's are recommended for a central control room installation. Give reasons for and against following the recommendation.

2. How is the information, input from a specific sensor over wires that are connected to specific input terminals, displayed in a specific graphic display on the central operator station CRT?

3. What considerations should be made to assure best visibility of CRT displays?

4. Sometimes wiring terminations are made to a separate cabinet (called a marshalling or termination cabinet) and cabled to the local control module rather than being connected directly to it. What are the advantages of this type of installation?

5. Some suppliers use one keyboard while others use two at the central control installation. Why is this?

UNIT 10
Digital Control Using Minicomputers

When digital computers were first used in processing plants in the early 1960's, they were imposing assemblies of equipment. In some installations there was a cabinet devoted to drum or core memory; another to the input/output information, stored on large reels of tape that looked like a Hollywood production; and still another large cabinet was devoted to the mainframe. This contained the main storage, the central processor unit, and the arithmetic and logic units. Today, solid state technology has shrunk the size and increased the power of computers so that a minicomputer comparable in processing power fits into a housing no bigger than the refrigerator in a domestic kitchen.

10-1 First Uses of Computers for Process Control

Using the term "early history" in the context of computers to refer to a period as recent as 25 years ago is indeed evidence of a rapidly changing technology. The first reports of use for process control from that period related to installations in utility power plant facilities and chemical process plants. Pulp and paper plants, steel-making plants, and facilities for many other industries followed with computer-controlled applications.

In the beginning, controlling processes from a large digital computer was not universally successful, for a number of reasons. The first attempts were at the beginning of a long, steep learning curve. Computers were used because processes were difficult to control. It turned out that there was also some difficulty in understanding how to use computers to do the controlling. Not only did the programs have to be written to tell the computer how to run a process, but other programs had to be written to tell the computer how to function as a controlling device: how to scan inputs and convert them into digital values that the computer could use; how to file the information away so that it was readily accessible; how to behave like a controller, developing output values based on deviation from set point; how to hold the output values until new ones had been computed; how to track process conditions and to output values to peripheral hardware; how to interface with recorders and controllers and manual/automatic stations; and how to converse with them so that control could pass to the computer from a peripheral controller and from the computer back to it, without upsetting the process.

Before the programming problems listed above could be tackled, programmers had to learn new ways to program. Computers had been applied successfully to business applications, where information was input on punched cards. For process control, real-time information had to be collected from measurements made at the process and input continuously in the form of analog signals. Data processing engineers were reluctant to relinquish their authority over computer applications to process engineers. Horror stories, possibly untrue, are still told of programmers who tried to write process control programs in business machine languages like COBOL.

Programming remains a constraint. Perhaps the most advanced application of computers to process control depends on making use of their ability to be programmed as mathematical models of a process. Even today, to develop satisfactory models requires special talents not readily available.

The early models of computers were not very reliable; consideration had to be given to using two complete computers, one to back up the other. The amount of backup to use in any type of control system is a function of the cost of production loss if the control equipment fails. For very critical applications the number of computers was increased to three, and data had to be identical in at least two of them to allow programs to run. This, of course, inflated the already high hardware and software costs astronomically. The combination of high cost and low reliability made use of computer-directed control difficult to justify.

Because of over-optimistic expectations by suppliers and users of their abilities to apply this new tool, design and programming costs often exceeded first estimates. Even today, programming costs can amount to two thirds of the entire system cost. On the early part of the learning curve, deliveries were extended and then extended again, primarily because programming took a lot longer than had been expected. Users became disenchanted, and application languished.

In spite of the fact that these setbacks caused users to lose their enthusiasm for computers as an answer to all their problems, the experience gained from many installations confirmed predictions that computers can be beneficial for process control in terms of increased throughput, higher quality, improved energy management, and greater profit. When microprocessors promoted the development of relatively low cost minicomputers, computer-directed control became a reality.

10-2 Using Mainframe Computers for Process Control

Today, the controlling functions of digital computers are being transferred to special-purpose computers distributed throughout the processing area, with less emphasis on direct control emanating from the big computer itself. Minicomputer application is being concentrated on computation and supervision of activities. When computers are used directly for control, however, there are some aspects of this application that are unique to digital computers. These will be reviewed briefly at this point.

Control systems programmed for computers use two types of control strategy; one is called direct digital control (DDC), the other, supervisory control.

Direct Digital Control

Direct digital control operates just the way it sounds. The computer looks at process variable values in relation to models of the process and of controller functions and computes the position that the control valve should take to bring about proper process performance. As a result of the computation, an output value is generated and sent directly to the final control element. A signal corresponding to position is fed back to confirm that the desired action has been taken.

Supervisory Control

Supervisory control, on the other hand, computes set point values on the basis of measurements of the process variables. Reference values of a number of interrelated loops may have to be altered simultaneously to coordinate changes in the process operation. If the computer is achieving its highest potential, these values will be developed as the result of an optimizing program. Computed set point values are sent to peripheral analog controllers which then produce control signals on the basis of deviation of the measured process variables from the computer-developed set points. The analog controllers may be pneumatic, operating to vary an air pressure from 3 to 15 psi; or they may be electronic, operating to vary an electric current from 4 to 20 milliamperes.

The process operator will still interface with a panel-mounted controller, but now the controller reference value comes from the computer. This analog operating station will have three modes of operation: local-automatic, computer-automatic, and manual/automatic. There must be a satisfactory way to transfer between the three modes to provide bumpless, balanceless transfer from local-automatic control to computer or backup modes without variation of the current output value. There should be provision so that the operator can manipulate the set point when the station is in local-automatic mode; or the valve when the station is in manual mode. In computer-automatic mode the computer will manipulate the set point. Therefore, all supervisory systems require that the process variable be connected to the analog station input and also to the computer input. Preferably, the computer should have the set point connected to it also, so that it can track the value created by the operator in local-automatic mode. Some programs require set point feedback changes to match the slope of the process input variable signal, even though the values do not have to be identical.

When the computer regulates the set point to provide control in a regular manner, the valve position has to be tracked. In manual, the set point may track the process variable and align the controller set point 1:1 with the process.

Supervisory control has reliability because there is an analog controller available to take over control, even if the computer fails. It is more expensive because the interfacing analog controller, in addition to its normal cost, must include components that convert the computer-derived output to an analog input. It must also have circuitry that permits communications to be established and maintained with the computer. DDC is less expensive, but the hazards in case of loss are great because all control loops are controlled from a single machine. As with all reliability considerations, the cost of providing assurance of continued production must be determined in the light of the costs incurred by failure.

10-3 Control Output Algorithms

When DDC is used for control, two different algorithms are used to develop the output signal sent to the final control device. One is called the position algorithm, the other the velocity algorithm.

The position algorithm computes a value that defines the position to which the final control device should move. Integration of error is performed in the computer. Some sort of position feedback device should be provided to confirm that the command has been carried out. The velocity algorithm, on the other hand, computes a value of rate of change at which the final control device is to move, or the distance it is to move in a specific time. Then it checks for deviation again and computes a new rate to correspond to the new condition. In effect, the final control device is used as an integrator.

The position algorithm requires initializing after every change so that other computed values, resulting from feedforward signals, for example, can be properly processed. This is not necessary when the velocity algorithm is used, because all computations are relative to the rate of change.

Both the velocity algorithm and the position algorithm are used for control from computers. The relative advantages and disadvantages are beyond the scope of this text but are clearly explained in Reference 1.

10-4 Input Sampling

A characteristic of digital control is that measurements are not made on a continuous basis but during a scan period. This means that the process may be changing between sample periods. Updating must be fast enough so that the shared time of the digital controller is as effective as the real-time control of the analog controller. This is a function of the loop and its dynamics. Flow loops may have to be updated every ½ second, while pressure and level may be satisfactory with 5-second updates and temperature with 20-second updates. Measurement of position requires sampling at intervals, depending on the velocity of the object whose position is being measured and the desired accuracy of measurement. Chemical composition rates must also be sampled at a rate appropriate to the rate of reaction; but there is no advantage in having sampling exceed the response time of the analytical device making the measurement, and this is usually, unfortunately, quite slow.

Scan cycle time is also a function of computer memory available to hold input information and the time consumed by computer commands used in the application programs. The effective sampled value is usually calculated by averaging the real-time value over the sampling period.

10-5 Output Signal Forms

The binary output signal that has been developed by the computer may be converted to a current output form, or to a string of pulses. The output value will be held at a constant value during each scan period, then updated and held at the new value until the next scan is completed. Current outputs are usually in the 4 to 20 mA range, and are applied to analog controllers as a voltage.

Trains of pulses are often used to change the set point values of pneumatic controllers. The velocity/position algorithm duality of output signals applies in this case. The pulse train signals will operate on stepping motors to move a set point potentiometer or a pressure regulator shaft to a specific position. Alternatively, signals operate to drive a variable speed motor, or to drive a fixed speed motor for a specific amount of time, so that a velocity-type signal is used to develop new set point values.

Each applied pulse of a computer output pulse train produces fixed output shaft revolutions. Steppers require special logic so that forward and reverse travel can be accomplished. Pulse counts from 500 to 2000 pulses for full scale travel are used; the most common is 1000 pulse full scale for 3 to 15 psi variation.

10-6 How Minicomputers Are Used

The evolution of the use of digital computers for process control has included the following stages:

- Data logging
- Data reduction and reporting
- Alarm monitoring and reporting
- Computer-adjusted set points (loops still analog)
- Direct computer control of the process
- Minicomputers used for report formatting, computation, optimization, and mathematical modeling to direct distributed process control equipment

The trend today is to use minicomputers as peripheral devices, leaving the routine operating functions of control to many special-purpose control-processing computers distributed throughout the system.

Used in this way, the minicomputer restores to the user the ability to formulate the functions of his instrumentation installation that satisfy his unique requirements, by writing application programs. This is a freedom from constraint denied to the user of digital control equipment packages defined by catalog numbers.

Application programs will be written using a high-level language. There is increased interest in newer languages like C, and LISP, but most programming will be done in BASIC or FORTRAN although PASCAL is frequently mentioned as a worthwhile alternate.

As is the case with all new acquisitions, the benefits of being able to create take getting used to. Engineers accustomed to analog equipment applications who are given digital equipment to work with for the first time will usually use it to duplicate their accustomed analog implementations. This is regrettable, because they are not making use of the power of their new tool. The minicomputer can be used to compute and, by doing so, makes use of many measured values that by themselves would be insignificant. Its potential should be limited only by the imagination of the person who applies it.

Computations that have been used as a basis for control include:

- Process modeling
- Optimizing
- Efficiency
- Energy consumption and allocation
- Fault analysis
- Batch scheduling and organization based on orders and delivery schedules

Computers may be used to perform control functions, generate graphic displays, log data, develop programs, and many other process-related activities. Another use of computers is operator monitoring. Operators can change set points, make computer/manual changes, and change tuning constants, but a record is kept by the computer. Alternatively, permission to make such changes can be locked out by an edit key, or given only on recognition of a password.

Other functions a computer can perform include changing gain of input amplifiers (programming sets the gain so that measurements have greater resolution at certain stages of process than at others); scanning for alarm status; generating data logs; performing calculational or sequential operation; and maintaining handshaking with interfaced peripherals.

Special control algorithms can be developed for special-process conditions. For example, if it is necessary to calculate a median value and do one thing if the median is exceeded by a process variable but another thing if it is not, a special algorithm can be written to do so. In fact, parameters in an algorithm can be varied to suit conditions, so that the algorithm is optimized for most productive use. This would be impossible, or at least very awkward and expensive, using analog equipment. Indeed, it cannot be done by the user of pre-programmed digital equipment such as distributed control systems.

Computers can be programmed to monitor process conditions and generate alarms when conditions are approaching an out-of-limit condition. They can then call up instructions for an operator from a table of instructions to tell him what to do to avoid a critical situation developing or to shut equipment down in an orderly fashion if the situation becomes critical. An intriguing use of computers is the generation of real-time models, but running at an accelerated rate to predict what may occur in the immediate future on the basis of current measurements and trends, generating messages to an operator to tell him to make changes before a situation develops that would be dangerous to personnel, equipment, and product. This permits anticipating upsets already in the making but not yet developed to the point that process values have begun to be affected.

In addition to direct control, computers are being proposed today as key parts of networks of hierarchical control. The computers used in these networks may interface over a data highway with other digital systems to accumulate data, process it, and communicate the information collected and generated to management levels where it can be used for decision making. They will also process commands received from the management level and transmit the results to departmental control systems.

REFERENCES

1. Theodore J. Williams, *The Use of Digital Computers in Process Control*, Research Triangle Park, NC: Instrument Society of America (1984).
2. Mellichamp, Duncan A., Editor, *Real-Time Computing*, Van Nostrand Reinhold Company (1983).

EXERCISES

1. How do computers used for process control differ from those used for business or scientific applications?

2. Values used in computations may be defined as integers, real (or floating point), or logical. Real values may be processed as single-precision or double-precision numbers. What do these terms mean?

3. What are interrupts, and how are they used?

4. What are the criteria for selection of a computer system to be used for process control?

5. To what uses can mathematical models be put?

UNIT 11
Personal Computers Used in Digital Control Systems

Within the limitations of memory capacity, time required to process data, and the ability to withstand the environment of plant operating conditions, personal computers provide an inexpensive microcomputer addition to digital control systems. The IBM PC™ and the Apple II™ are the most commonly used at the present time (1986) but Kaypro™, Compaq™, TRS-80™, Corona™, AT&T Model 6300™, TI Professional™, Columbia Data™, Heathkit Model 151™, and Televideo Data™ computers are listed by interface board suppliers for compatibility with process-oriented hardware and software. More and more software is being written to interface process equipment and measurements with the computers and to perform industrial-process-related applications. This is particulary true in the case of the IBM PC™, XT™, and AT™ personal computers.

11-1 What Is the Personal Computer?

Because microprocessors and large scale integrated circuitry have reduced the cost of digital circuits, minicomputers with a lot of memory capacity and powerful operating systems have become available in the $2,000 to $10,000 range. Since this is the same range as the price of automobiles, a personal computer is no more expensive than a second car. Whether the average person has applications that justify the expense is a moot question, but more and more industrial operations personnel are finding that they do have use for this scaled-down minicomputer.

The physical package called a personal computer consists of a power supply, with fans to remove the heat generated by it, packaged with printed circuit cards containing an operating system designed around a microprocessor. A keyboard, built into the housing or connected by a short cable will complete the basic package. Auxiliary devices may include a CRT display screen monitor, some sort of bulk memory storage device, and a printer. Communications processing and a modem will be needed for networks interfacing.

The most common auxiliaries will be a disc drive and a printer. These will be connected physically to the computer housing through cables and plug connectors. Electrically, they will interface through software ports, selected by programming. Early personal computer models used a tape cassette for bulk memory storage but the floppy disk, with a magnetic surface coating, soon took its place because of greater storage capacity and much greater speed and accessibility to data. Computers can access one or more (usually up to four, sometimes eight) disk drives. Disk drives will be for floppy disks, usually 5¼ inches in diameter (although a 3½ inch disk is beginning to be used), or hard (permanent) disk drives. Externally mounted 8-inch disk drives can also be used.

The 5¼ inch disks will probably be double sided, double density, with storage space for 320K, or more, bytes. The hard disk storage is much larger, holding 10 to 30 megabytes of information.

A printed circuit card will hold the most important parts of the basic package. It will contain memory, one or more microprocessor chips, a clock, and the registers, encoders, decoders, and all the other components necessary to handle information flow. 64K bytes of memory is usually supplied as standard. More can be added by plugging in expansion cards internally or external to the computer in separate housings.

There will probably be at least two printed circuit boards in the computer in addition to the one containing the operating system and basic memory. One will be a card serving as a disk driver and another will be for communication with the CRT display and the printer. Additional cards can be added, using the expansion slots, to provide graphics capabilities for the display, communications to external devices, and additional memory capacity.

11-2 Operating Systems

Everything else exists as software, usually stored on disks. The most important software is the operating system program. The operating system is the manager of the computer's resources; it contains the library of software routines that determine how information will be moved about to perform the computer's basic functions. All of these routines relate a physical action — pressing a keyboard key, for example — to an informational action (for example, storing into memory the data value or address for the symbol for which the key is coded).

There are three major parts to an operating system. The most basic controls input and output transfer of information between the keyboard and computer, and between the monitor and the computer. The number of columns displayed, cursor control, and the routines for receiving a keystroke and producing a display from it are designed into the operating system. Many of these internal operations are interrupt-driven. An interrupt stops the application program after the completion of the command being executed at the instant the interrupt occurred, stores the address of the next command with all the important information that will be needed when the application program resumes, then checks a priority listing to see if any higher priority interrupts are waiting to be serviced. If not, the program goes off in another direction as directed by a pointer that the interrupt activated. When the branch program is completed, the interrupted application program resumes, unless another interrupt has occurred, and restores all the stored data.

The second part of the operating system supervises the disk system to service application programs on the disks. The routines at this level can open files, for example, and read and write to the disk. The most important part of the operating system, however, contains the programming for the microprocessor, the executive programs that include the logic necessary to find and run programs input from the disk systems.

The operating system in most of the personal computers used for industrial applications is MS-DOS™, written by Microsoft, Inc. to use the Intel 8086®, a 16-bit microprocessor. Earlier 8-bit processors such as the Zilog Inc. Z-80® chip used an operating system called CPM, and an updated version of that system, CPM/86™, can be used with the 8086 chip. PC-DOS™ is IBM's version of MS-DOS™. There are small but important differences between PC-DOS and MS-DOS, but most computers operating with one are compatible with the other.

11-3 How Are Personal Computers Used for Digital Control Applications?

The personal computer can be of benefit in a laboratory or an industrial plant operating department, even if it is used only as a piece of office equipment. Commercially available programs that can be purchased by the user on disks can be helpful to the manager, engineer, or technician. Word processor programs can be used for report writing. Spreadsheets can be used for organizing information. Application programs specific to departmental activities may be written by the user and saved on disks. It is possible to combine these programs with commercially available programs. Application programs can be written for analyzing data, for calculating operating values, for keeping inventory of spare parts, for developing maintenance schedules, and for assistance in process design. Through modems, developed information can be sent to other computers, and information can be received. In other words, the same functions available to the owner of a personal computer are available to the industrial manager, and his applications are so many times greater that it is easy to justify the expense of a computer, if only for office use.

If the user is going to write programs, there must be software that interprets from the language in which the program is written to the machine language of the computer. Most commonly, an interpreter program for some form of BASIC will be supplied on the disk containing the operating system. Compilers have been written for Pascal, FORTRAN, and other high level languages. Assemblers are available for translating mnemonics that make up source programs written in assembly language into object programs that can be used by the machine. As part of the original system software there should also be some sort of text editor program, so that the programmer can format his program to match the compiler's syntax and grammar.

Used as a desk top device, the computer is changed from one task to another by changing disks. It might be used by an operating engineer, for example, to accumulate process data and store it in working memory. He can then change disks and process the data to calculate operating ratios and productivity, rates of reaction and percentage completion; change disks to a word processing program and prepare a report using the calculated data, then print it in a specific format; and finally go to another disk that accesses different external equipment than the first one mentioned, to change set points and operating parameters and initiate another batch process operation.

To run a program, the user activates the operating system, inserts the disk on which the program is stored into one of the disk drives, and enters its name from the keyboard. The operating system will load the program from the disk into a RAM working space in its memory, and perform the information transfers determined by the program.

Programs that have been developed plantwise can be accumulated and maintained in a common data base file management system, enhancing the plant process design and operating capabilities. Software programs can access the stored programs and use them in conjunction with real-time process data; or computers can use programs during off-line usage.

An information network can be installed to provide access to plant, laboratory, and pilot plant data from remote locations. Intelligent recording of a plant's procedures can increase the potential of any production unit when the programs of all units are universally available. Portability of programs by transferring discs generated in one department to the personal computers located in other departments is another way to make the information sources of one group accessible to others who may have use for it.

11-4 Real-Time Use — Interfacing the Process

These activities use a very small portion of the powers of the personal computer. The next step is to incorporate real-time processing. This requires interfacing so that information, in addition to entry from keyboard and disk, can also come from the process and, in addition to going to the screen, disk, and printer, can be sent out to the process. Used as a dedicated process control or monitoring device, the personal computer will operate in conjunction with specific process equipment, interfacing through specific driver programs, and will perform only the functions programmed. This will be the case when the application is process oriented, particularly for processes that must be monitored continuously. In most cases, however, output values will be held constant if the program is removed, so that even for dedicated applications the personal computer can be used off-line for report generation and other activities.

To support real-world applications, hardware must be added and software must be written. Hardware and software are commercially available.

11-5 Hardware Enhancements for Process Application

In order to get information from the process, some form of data acquisition system must exist. The process signals may be either discrete on-off signals or analog values. They must be terminated, and the analog inputs must be multiplexed, conditioned by filtering, amplified with proportional gain and bias, isolated with sample and hold circuits, and changed to digital form by a digital-to-analog conversion. Under the supervision of a driver program, this information must be brought into the system and stored in designated memory locations. After the information has been processed by the application program, computed values that are going to be used as process commands must be stored in memory, then sent out under the supervision of the driver program. Discrete commands can be routed through TTL circuitry, while commands that are going to be used to regulate valves and other proportional control devices must be changed from digital to analog values.

A driver program supervises all the transactions that take place between a device and the programs that call it. It makes the transfer of data transparent, so that an external device appears to the operating system to be just another port. It will translate between the language of the computer operating system and that of the device. It may include handshaking routines to announce the presence of information to be transferred, or the need for it; checking routines to test for accurate transfer; and routines to transfer blocks of information. It is the program that contains the codes for asking for information from the appropriate part of the data base of the computer, and for storing it in buffers that can be accessed by the external operating system. It is usually brought into operation by a CALL (or IN or OUT) command from the application program.

There are two ways that hardware can be added to make the personal computer function as a process or laboratory control device. One is to use expansion slots in the computer printed circuit cards to mount the data acquisition and processing equipment; the other is to mount the data acquisition hardware externally. The Apple II™ computer has eight expansion slots, but all may not be available. One at least will be used by a disk controller card. The IBM PC™ has five expansion slots, and again, not all will be available. One will be used for disk drive control and another for the monitor and printer drivers. If the system has the capability to display graphics and color, if memory has been expanded above 256K bytes, and if there is a modem for external communication, the remaining three slots will have been used. IBM offers an expansion chassis with eight additional expansion slots, connected by cable to the operating system bus.

The alternate solution is to mount all the hardware externally and bring the information into the computer through a software port under supervision of a driver program. This allows more leeway in the size of the system and the speed of processing. Software ports are included as part of the operating system. One of them will be addressed for each of the disk drives: one for the keyboard, one for the CRT screen display, and so on. As far as a port is concerned, any external connection is just another device to send information to and receive data from. Usually there will be a number available (the IBM PC has 256), and so no changes have to be made to the basic architechture of the computer.

A protocol must be established for the information transfer. RS-232C is the standard used most generally. Hewlett-Packard® and others use the IEEE 488 interface. The data can be brought in one item at a time and sent to specific addresses, but this is uneconomical because the operating system has to be involved. It will have to know when a conversion is taking place and either wait for completion before doing anything else or keep coming back (polling) to find out when another conversion can be initiated. An alternative is to have an interrupt at the completion of conversion notify the operating system, but this uses more memory and time. More commonly, entire blocks of information will be transferred by a direct memory access (DMA) procedure into an area set aside (mapped) for it, bypassing the operating system so that there is no wait for interrupts. DMA permits logic associated with the external circuitry to have direct access to memory with operating system intervention, so that data transfer does not have to wait for word-by-word transfer. An entire block of data can be transferred into a mapped area, starting at a designated address and indexing addresses for each byte while simultaneously counting down the total number of bytes. Transfer speeds as high as 200,000 samples per second are listed.

A sampling of catalogs for suppliers of hardware for process control systems includes the following expansion boards:

- A/D, D/A; 50,000 samples/second, 16 single-ended analog inputs, 2 analog outputs, 4 digital inputs, 4 digital outputs
- Graphics processor boards
- Line drivers for instrumentation-to-data acquisition systems
- Timer and stepper interface
- Additional memory; clock and serial port. Boards and chips can be added to the IBM PC to bring the memory up to 1024K bytes

Signal conditioning boards are available for the following inputs:

- Voltages, millivoltages, strain gages, resistance thermocouples, frequency, and discrete digital signals

Boards are available for output signal generation for the following regulating devices:

- Voltage, current (4 to 20 mA), triac, pulse, and discrete digital signals

11-6 Software for Industrial Applications

Application programs, making use of the information that has been collected and brought in, can be written or purchased. The IBM PC and the Apple computers are the most commonly referenced for commercially available digital control applications because the majority of the software written for application programs runs on these two types of computers. Programmers will write software for the supplier who they believe has the greatest potential for marketing their product; therefore, available programs are concentrated around these two leaders in the marketplace. Some suppliers of single-loop digital controllers supply programs that allow their product to be supervised by a personal computer.

Many application programs can be used in conjunction with business-oriented programs like Lotus 1-2-3™, Symphony™, or Visicalc™, displaying the collected data information on spreadsheets and in graphic displays. Application programs can be written by the user that can exchange information between unrelated programs, using DIF (data information format). A file saved in this format is a text file that can be read by other DIF-supporting programs.

Application software is commercially available to support the following applications:

- Proportional, integral, and differential (PID) control
- Event alarms
- Data trending
- Data logging
- Intelligent modems, to transmit information to other computers
- Polynomial analysis
- Curve fitting
- Error analysis
- Statistical analysis
- Computer-aided drafting (CAD)

Many of these are menu-driven and require very little or no programming experience by the user.

11-7 Performance of Multiple Tasks

Practical industrial applications demand that simultaneous activities be attended to, a requirement of operating systems called "multi-tasking". At the present time, operating systems used by personal computers will operate only one program at a time. The trend in the development of personal computers used for industrial applications is toward modifications that permit multi-tasking.

A number of operating systems used for larger minicomputers can perform multiple tasks and have been proposed as replacements for the PC-DOS system. The name most often encountered is UNIX™, a "time-shared" system licensed by AT & T, its originator. Venix/86™, from Ventura Corp., is a version of UNIX with a number of real-time extensions for measurement, control, and networking. Other systems are VRTX™, from Hunter and Ready: systems modified from Intel's iRMX™, a real-time operating system for the 8086 and 80186 microprocessors; IC-DOS™, another UNIX-like system; and VME™ and Multibus™, designed for larger machines. At the time of this writing UNIX is frequently mentioned as the most likely replacement.

This does not mean that only one control loop can be supervised from one program. A program can be written to control many loops. One way to write it is to have each control sequence described by a series of commands organized as a subroutine, and have the main program GOSUB them in any order desired. Another possibility is to use interrupts to implement branching to subroutines.

Time sharing is another form of multi-tasking. An interval of time split between a number of uses allows separate tasks to be performed in the splinter of time allotted to each one.

Systems have been developed that permit running two separate programs, one in foreground, the other in background. Data acquisition, processing alarms, and process control with interactive graphics can be selected through the use of menus and interwoven with the main operating program without interfering with it.

All of these forms of multi-tasking are transparent to the user, unless there are so many tasks, or branches, or subroutines called for that there is an appreciable delay in repeating the processing of any specific one of them. When this happens, display updating will be noticeably slow. Of course, there must be enough memory to handle all the operations required.

11-8 Applications

The personal computer may be used as a desk top device, or it may be dedicated for use as a process controller or monitor. As examples of the uses to which personal computers are being put, the following specific applications have been taken from some of the referenced articles.

1. The most extensive use reported is for data acquisition, either for monitoring or in conjunction with other standard programs like Visicalc™, Multiplan™, or Lotus 1-2-3™. External devices are reported to be able to handle as many as 100 input/output signals, expandable to 1000 by interconnecting and multiplexing remote units.

Process information inputs (they could be vibration analyzer signals from test stands, chromatograph values from a fractionating column, or levels measurements from a tank farm) are scanned, the measurements digitized, parameters checked against limits, and the information presented on a color monitor in bar chart form or printed as a report.

2. Another use gaining acceptance is for programming programmable controllers. Personal computers are also being used to program robots, a related application.

It has been suggested that personal computers will eventually take the place of programming modules presently supplied by PLC suppliers, who will supply software on disk for the user to load into his personal computer. Advantages of the personal computer over the PLC programming modules are that they have more memory; they have built-in editing and file management software that make it easier to enter, edit, and store programs; and the word processing support from word processor disks encourages documentation, generation of manuals, and keeping of records.

Personal computers are not being used for the fast, event-driven applications for which programmable logic controllers have been developed because they are not, in general, fast enough to be dependable when sequential timing is a requirement.

3. Process modeling applications use data collected from an operating process to simulate operating upsets and emergency conditions. Depending on the computer used, high resolution graphics are available for displaying the process on the CRT monitor. This can be a useful training aid, a tool for designing control and operating systems, or even a means for predicting emergencies before they take place based on accelerated computations using real-time data. This sort of alarm system would alert the operator to a condition that would occur if corrective action were not taken, and at the same time display a procedure to follow to avoid the emergency.

4. Used with weigh stations, personal computers can access data from individual stations, use it to compute standard deviation or make other statistical analyses, and produce quality and production reports formatted to user standards.

5. Personal computers are being used for cycling pneumatically controlled or timed activities used in collecting data, or in conducting batch chemical operations. This last application is of interest when many recipes or sequences of ingredient additions and reaction times are required, because of the ability to store the information on disk and use it as required.

6. Creating trended records and displays to preserve plant activity histories; grouping of plant variables for interactive spreadsheet analysis; data base organization for corporate and marketing planning; maintenance overhaul, inventory records, and production scheduling; and verification of calibration are other uses.

7. Combined with digitizers, personal computers are being used for computer-aided design, allowing an operator to create and alter drawings.

8. The supervisory control of a number of analog process controllers, interconnected by a data highway, can be programmed by personal computers. Networks of them can be used with process controllers in this way to provide a central control room operator with displays that show operating parameters of any individual controller, and the capability of changing control parameters. This application will be most successful when the processes for which it is used do not change rapidly as, for example, temperature control. However, there are reports of successful application to flow and pressure control. Since this is supervisory control, the actual control functions are performed by conventional analog controllers, not by the computer.

9. The greatest amount of commercially available software at the present time is directed toward biological and chemical laboratory and pilot plant applications. Ideally, a laboratory project will monitor large amounts of data, collected while the parameters of the equipment and process are changed under carefully controlled conditions over specific operating ranges. The data collected is analyzed, curves are fitted to it, and reports are issued. The front-end and back-end hardware exists to get data into the computer and to get control signals out of it. Programs also exist that can call other programs to collect data and then return to a menu-selected procedure for processing it. Software for the data management requirements specific to scientific computing has been written and is commercially available to perform the following tasks:

- Data can be stored and displayed in spreadsheet formats to suit the column and row arrangements best suited for analysis or reporting. Standard spreadsheet software used for office and business applications (Lotus 1-2-3™ and Visicalc™, for example) can be accessed in conjunction with the special programs under the supervision of a master program written in BASIC, Pascal, or another high level language. The next step is to extend the application to include budget preparation.
- Data can be edited, sorted, searched, manipulated, and arranged in subsets by subject, as well as time and date.
- Graphics can be prepared: data can be presented as bar graphs, pie-charts, scatter plots, histograms, and three dimensional plots.

- Linear and nonlinear regression analysis can be performed to fit the data to curves, which can be displayed. Using the display screen, much of the curve fitting can be done by trial and error, much as square roots can be computed by iteration with a hand-held calculator that does not have a square root function, homing in on the answer.
- Statistical analyses can be performed. Statistical tools available in software include computation of standard deviation, many types of means with many ways to calculate them, distribution functions, correlations, and random number generation. Extensions of this application are fault analysis and determinaton of causes for off-spec production.
- Some software allows the development of analytical and production models, with which the collected data can be used to generate hypothetical process conditions further down the production line.
- Final reports can be produced that combine data, graphs, tables, and text into finished form presentations. Some software includes built-in word processing capabilities; others allows interaction with standard software word processing packages.

Practical use of these programs for personal computers is being made in process plants as well as in laboratories. The following examples are listed:

- Analyzing processes on line, to identify the causes of out-of-control conditions and to simulate, using process models, possible corrective action.
- Analysis of chromatograph outputs including integration of areas and display of gradient profiles, with enlargement of any portion for detailed inspection and analyses. These include area %, normalized area %, concentration analysis with automatic concentration of sample weight and dilution factors, and peak detection. Outputs can control external devices such as autosamplers, valves, heating baths, etc.

It must be understood that all the tasks listed above can be programmed into a minicomputer. The utility of a personal computer is limited by its memory capability, and there must be a choice between saving data on many floppy disks or on a hard disk. Where the software exists to let the personal computer do the job, it gives significant cost savings over the use of the minicomputer.

11-9 Limitations

It has been said by enthusiastic supporters that use of personal computers puts the decision-making capability for process control in the hands of the operating engineer. It is easy to think that the personal computer will do everything after reading about all the advances that have been made in its use. It won't do everything. In addition to the constraints imposed by single-tasking operating systems, already discussed, there are limitations in the size of memory and there are very severe environmental limitations.

Memory size limitations include internal memory on chips, up to 1024K bytes; disk memory (using double density, double sided disks), up to 368K bytes; and hard disk memory, up to 32 megabytes. When requirements exceed these limits it is time to think about using one of the larger machines.

Personal computers are also limited to locations free from vibration, high temperature, and dirt, although rugged models are beginning to appear. Personal computers packaged for use in the factory production areas were exhibited at the 13th International Programmable Controllers Conference and Equipment Display at Houston, Texas, in April 1984. In that year, IBM introduced the IBM 5531™. This model is functionally identical to the IBM PC™, but offers extended protection against vibration, dust, temperature extremes, and voltage transients. Still newer (third quarter, 1985) are the IBM 7531™ (for floor mounting) and the 7532™ model (for rack mounting), 80286-based models (32-bit operating systems) functionally identical to the IBM PC/AT. Both units offer extended protection against temperature extremes, vibration, and shock.

Using the PC-DOS™ operating system, personal computers that are commercially available do well for data acquisition and data processing jobs. Hardware and software is commercially available for performing PID control functions, but the single-tasking personal computer has limitations that inhibit performing real-time process control directly. The operating system may be too slow for some applications. It is involved with controlling processor executive programs, I/O to terminals, disk drive and printer access. It has to handle file access operations, and read and write instructions to disk memory. It has to download from disk memory to the RAM-resident memory. All this activity means that it cannot be concentrating on process control.

Improvements that will upgrade personal computer design for real-time process control are being made. They include the following:

- Using bubble memory. This protects against memory loss in case of power failures, takes the place of disk memory, and allows much faster access than disk memory.
- Using hard disks instead of floppy disks. This increases the amount of memory and speeds up data transfers.
- Using portions of internal memory for application-program storage (RAM disk). Data transfer between these areas and the operating system is much faster than from floppy disk memory.
- Modularizing the functions presently performed by a single motherboard in the IBM PC.
- Modifying design to provide capability for multi-tasking.

In spite of its limitations, the personal computer has assumed a commanding place as a digital control device. This has come about because it is relatively cheap, it allows the user to program to suit his unique requirements, and it can be interfaced to very versatile data base, spreadsheet, and word processing systems. As its capabilities are improved its price will increase, but it will still be a bargain because of the personal control capability that will remain.

REFERENCES

1. Merritt, Rick, and Terry Persun, "Personal Computers Move into Control," *I&CS*, June, 1983, pp. 39–44.
2. Basta, Nicholas, "Use of Personal Computers Growing among Engineers," *Chemical Engineering*, September 19, 1983, pp. 14–17.
3. Bailey, S. J., "Desktop to Plant Floor — Personal Computers Can Help the Control Engineer," *Control Engineering*, December, 1983, pp. 49–54.
4. "Personal Computers: Frugal Path to Specialized Control Systems," *Control Engineering*, July, 1984, pp. 89–93.
5. Cleaveland, Peter, "What's New in Data Acquisition?" *I&CS*, May, 1984, pp. 29–33.
6. Miller, Timothy J., "Data Acquisition and Control with Personal Computers — A Blossoming," *Control Engineering*, May, 1984, pp. 81–82.
7. Duncan, Ray, "PC-DOS versus CP/M-86," *Softalk for the IBM PC*, October, 1982, pp. 38–41.
8. McQuaid, Jim, "Update on Personal Computer Operating Systems," *I&CS*, November, 1984, pp. 57–60.
9. Bailey, S. J., "Personal Computers 1985: Pawns in the New Micro Game of Value Added Control," *Control Engineering*, July, 1985, pp. 66–69.

EXERCISES

1. How can you protect a personal computer installation from public access?

2. How can you protect a program from being accessed or copied?

3. What are the functions of fans and filters in computer design? What maintenance steps should be taken to assure their continued functioning?

4. What commercially available software is available for use in conjunction with application programs for process and laboratory applications?

5. What types of printers are available for use with personal computers?

UNIT 12
Process Applications

12-1 Introduction

Microprocessors have made new tools available to the process industries — tools with new capabilities. Data acquisition and loggers, personal minicomputers, programmable logic controllers, distributed control, single-loop and multiloop controllers are permanent categories of digital instrumentation. The venerable strip chart recorder now is supplied with printed circuit boards, a communication port, and a keyboard that can be used to program ranges, alarms, printing speeds and sequences that can be changed only if one knows the proper code word.

All of the categories of digital instrumentation discussed in Units 6 through 11 are being used because they have enhanced abilities for performing applications peculiar to the needs of process industries. The special capabilities provided by digital control devices that make them outstanding for applications in industrial process control include the following:

(1) *Control from multiple recipes.* It is important to batch operation control to be able to change values for set points, tuning constants, heating times and rates, flow rates and quantities, reaction times, and soak times quickly and with complete repeatability, so that as product sizes and compositions change there is reproducibility of control performance. Production of a product at different times while maintaining identical conditions, with consistent quality of product, is vital to quality control. Productivity increases because setup time is reduced when the control parameters are set up in software.

(2) *Sequential operations.* It is important to be able to implement and repeat a multistep procedure based on time and event occurrence, with the elimination of human error and manpower reduction.

(3) *Optimizing.* Best products and profits result from operating under conditions that produce the greatest profit for the least cost. This is particularly difficult to achieve when several interrelated variables change during operation. Sometimes these conditions can be determined accurately by depending on operator intuition, but only in the simplest cases.

(4) *Collection of quantities of data, with statistical and out-of-limit analysis.* This is the best way to detect incipient off-spec operations before they occur, determine and document reasons for off-grade products, monitor equipment operation, and analyze production losses from records of conditions before and during upsets.

(5) *Specialized report formats.* Only reports of computed as well as measured values can provide management with information suitable for making decisions.

6) *Management decision making.* Digital instrumentation provides help in scheduling and inventory control when plant throughput changes frequently because of orders, processing rates and composition changes, and equipment availability.

(7) *Centralized control areas.* Operation from one point for interrelated processing areas, or for multiple machining and heat-treating installations dispersed through one or several shops helps to prevent mistakes and reduces costly inventory. The use of data highways reduces installation costs.

Increased productivity, reduced cost, and improved product quality result from all of these attributes of digital control.

There is another reason that the use of digital control devices will increase. This is the capability of a computer to guide and supervise their functions over a communications network. Underlying all control strategy is the great influence that interrelations of all controlled variables have with each other as well as with the ambience of production and economic constraints in which they must function. Expanding the concept of a control system from a single piece of equipment to an entire production facility, it becomes obvious that analog loops simply are not enough to do the job. They can control individual parts, but not the whole entity. Digital equipment, through the programming of supervisory minicomputers in conjunction with the flow of information between devices, can control the whole; this is the ultimate promise of digital instrumentation and its significant contribution to the progress of instrumentation.

Units 12 and 13 will present examples of applications of the categories of digital control devices discussed in the preceding units. Each type of digital control device can do some of the functions that can be done by other types. Data acquisition devices collect data, but they can also process the data and perform computations with it. The majority of applications of programmable controllers are for operation of discrete control systems, but these devices also have continuous process control capabilities. Distributed control systems have powerful algorithms designed to be used for continuous process control, but they also have logical decision algorithms that allow application for batch control. Minicomputers can be programmed to perform almost any combination of functions and are most useful in performing sophisticated computations, but they can also implement direct digital control. There is much overlap, if the application is considered in the context of what an instrument can do.

Consequently, to make the examples more meaningful, they are process-oriented rather than hardware-oriented. The strengths of each category of digital control device will be evident from the uses to which it can be put. The pattern that the applications examples will support is one of gathering many separate influences into one coordinated whole. Using digital control devices provides the strength to make an entire process, or even an entire plant, a closed-loop feedback control system where the set point is profit margin. Analog instruments can perform the individual single-loop activities — not as well as digital devices, perhaps — but only the digital implementation can put it all together.

12-2 Data Logging and Processing Applications

Data acquisition is particularly useful when large amounts of information must be collected, and even more so when the information must be analyzed, processed, and reported. A number of applications were listed in Unit 6, and several will be discussed in more detail in this section.

Tank Farm Inventory

Keeping track of the inventory in a tank farm must be done on a daily basis, or more frequently if inflow and outflow are constantly changing. The variable measured is tank level. Floats are the most commonly used sensors for determining levels in large tanks, although capacitance probes are used in some cases. A number of electronic devices can be used for transducing float position, including synchros, potentiometers, and brush encoders. A digital transmitter (Reference 1) communicating through an RS-422 interface is being used to send HDLC coded signals a distance of up to ten miles for this purpose.

Level is an inferential parameter; the variable of real interest is the gallonage of a specific liquid, not the number of cubic feet it occupies in the tank. Gallonage, however, cannot be measured directly.

A problem with using level as an indication of quantity is that while the level measurement may be very accurate, it may not truly represent tank volume. If the bottom is conical or the ends are dished, the level/volume relationship is not linear. Horizontal tanks are not always parallel to the ground. Of equal seriousness is the problem that, even if the tank is vertical and square or cylindrical, the density of the liquid in the tank must be known in order to make a weight determination of the contents. Density varies with composition and with temperature.

If composition, temperature, and physical dimensions of a tank are known, the level measurement can be compensated by calculation to compute a true weight or volume figure. A temperature measurement, or several that can be averaged, is required. In the case of the digital transmitter mentioned above, transmitted signals corresponding to level and temperature are processed by a microprocessor-based (Motorola 68000™, 16-bit) master control unit, with plenty of memory capacity for storing tables of tank dimension values. Transmitted analog signals can be converted to density-compensated digital values by numerous types of data acquisition devices with data processing capabilities.

Information desirable for display and reporting could include:

- Level
- Temperature
- Tank volume
- Contents weight
- Volume available for storage
- Status (filling, empty, full)
- History for a period of time
- Alarm indication (overflow, leaking, pump malfunction)

Temperature Profiles

Ceramic bricks are made slowly, moving on conveyors through long furnaces or kilns, past many firing stations that heat the inner surface of the kiln to provide heat for curing the bricks as they go by. There is no way to test the condition of the product in the kiln until it is removed at the discharge end. Only temperature measurements can give some assurance that the ambient conditions are correct for the operation needed.

A typical kiln will have three zones, with twenty thermocouples per zone. Firing rates can be established on the basis of data acquired by measurement, averaged and processed statistically to determine the consistency of the firing process.

Similar measurements must be made for rotating kilns used for cement, lime, and alumina, as well as for dryers. Each process has special conditions. Cement kilns use a curtain of chains to transfer heat from the combustion gases that sweep through the kiln from the firing zone to the stack. A differential temperature measurement across the chain section is helpful to preserve its life. Other measurements are necessary at various "hot" and intermediate zones. Temperatures inside the rotating kiln are measured by thermocouples in ceramic sleeves or wells inserted through the kiln shell and brick lining. Since the kilns are rotating, the electrical signals from the thermocouples are collected on a trolley wire that encircles the kiln. Pantographs used to be employed to pick the signal off the wire and transfer it to extension leadwire for transmission to recorders, but today the receiving antenna of a stationary radio transmitter is used.

Typical of the capabilities of digital techniques, one such model consists of a 10-channel scanner, with six channels assigned to data and four used for calibration, synchronization, transmitter temperature, and zero reference. The last data pulse of each transmission is used to synchronize data between the transmitter and the receiver, with data refreshed at the scanning rate of approximately 310 Hz. Sample and hold circuitry maintains the data between updates.

Textile Thread Extrusion

Thread for textiles made from chemical feedstock (nylon, rayon, or polyester, for example) is produced by extruding heated plastic material through spinnerettes, tiny openings in spinning heads. The thread is made in huge air conditioned rooms filled with humming machinery, with literally thousands of spinnerettes operating at once. The chemical feedstock is fed into a melting hopper from which a feed pump forces the plastic into the spinnerette. Thermocouples are used to measure the temperature of the melted plastic, while the level of the plastic in the feed bin is determined with a resistance measuring device. The feed pump speed is constantly monitored to ensure proper feed through the spinner head and to detect a clogged head. All the measurements can be transmitted to data acquisition devices for logging, and in some systems a PID control algorithm is programmed into the data device to control the speed of the pump, based on a motor speed set point.

Incipient Failure Detection

A significant benefit of monitoring data is the detection of failures before they occur. This can be done by setting limits and generating alarms if the limits are exceeded. Some examples of the use of data acquisition equipment for this purpose follow.

1. In many nuclear reactors used for power generation, borated water is used as the emergency cooling medium. If the water temperature drops too low the borated solution will crystallize and clog the pipes, causing failure in the emergency cooling system. When this happens the system must be flushed and the solution replaced. If the temperature rises higher than necessary, deterioration of seals and packing increases. Alarm set points are used to warn when the coolant temperature is out of the acceptable zone.

2. Turbine fan mounts, aligned to eliminate torsional stresses, may be temperature-sensitive. Stresses developed on the turbine shaft can cause expensive failures. A method for avoiding such failures consists of monitoring temperatures and stresses and checking allowable limits as the fan heats up to equilibrium, then monitoring at equilibrium conditions.

3. Failure, with significant damage, can occur in large multistage compressors when valves stick. Measurement of valve temperature, with a check against limits and against the temperatures of all other valves, can be used to determine when one valve is deviating substantially in performance from the others, an indication of incipient failure. Alarms, with contact outputs, can be used to shut the compressor down if a potential failure is indicated. Compressor inlet and outlet pressures, oil pressures and temperatures, and bearing temperatures are other operating variables that should be monitored.

4. Monitoring temperatures in and out of water jackets around electric arc furnaces is desirable to prevent potential furnace wall melt-through if a jacket plugs up or leaks. The same precaution can be taken for the cooling jackets around tuyeres of blast furnaces.

5. Monitoring and controlling the heat trace wrappings around oil, chemical, or food processing piping, where movement of viscous material depends on maintaining process temperatures, is an important factor in preventing line plugging if the tracing wrappings fail to do their job.

12-3 Applications for Single-Loop Controllers

Many applications for which analog controllers were used when they were the only means of obtaining control were developed by combining measurements and functions to generate values that could not be measured directly. Percentage of reaction completed, Btu's consumed or generated, and mass flow are examples of desirable variables that cannot be measured directly and are usually inferred from other variables that can, such as temperature and pressure. Some of the combinations have been given familiar names, as, for example, three-element feedwater control (used for boiler drum level), internal reflux (used for end-product control in fractionation), and cross limiting (used in combustion control).

The same combinations of functions and variables can be obtained in distributed control systems that break algorithms down into basic functions (see Unit 9, Section 9-3). Some single-loop controllers have the flexibility of programming to provide this sophistication, far beyond the capability of the analog PID controller but at the same relative cost level.

Programmable digital single-loop controllers have software that the user can format to perform this sort of computation. This puts a tool into the hands of the instrument engineer that was never available when conventional analog control was used.

Fischer and Porter's Micro-DCI™ Series is used as an example of such a programmable microprocessor-based controller because of the number of process parameters that have been derived from it by programming. The basic controller can combine 59 functions, and a separate module, aptly named Chameleon™, can put together up to 400 of the various operands and operators that are valid for that instrument to allow an imaginative user to create many combinations. Basic operations with positive integers and floating point numbers include addition, subtraction, multiplication, division, comparing, swapping, biasing, and scaling; digital AND, OR, exclusive OR, and NOT, with conditional statements allowing jumping and branching; and several forms of PID, totalizing, and square root. Log to the base 2 is useful in chemical process rate of reaction computations, and variables raised to a power allow computation of polynomials.

Figure 12-1 illustrates a controller module programmed for ratio control. This type of control is used when the value of one variable must be maintained in constant proportion to the value of an uncontrolled variable. In the example, limits for both flows are imposed so that unreasonable operating conditions are prevented. The program uses less than 59 elements to go through a program that does the following steps.

1. Check the wild flow value. If it is less than an allowable limit, give it the value of that limit.

2. Check the controlled flow output value. If it is below its low limit, make the value that limit.

3. Compute the controlled flow value from the ratio. Check to see if the computed value is higher than a limit. Put out an alarm if it is, and make the set point for the PID the high limit value; otherwise, make it the computed value.

4. Perform the controller action.

Programming is done in a language that is designed to be used exclusively with this equipment. It is more complicated to use than fill-in-the-blanks programming because there is a form of coding to be learned, but it is much simpler than the standard high level languages because it is not necessary to make alternative choices depending on the way the commands are being used. There are a fixed number of operands, and when they have been learned the coding is easily remembered.

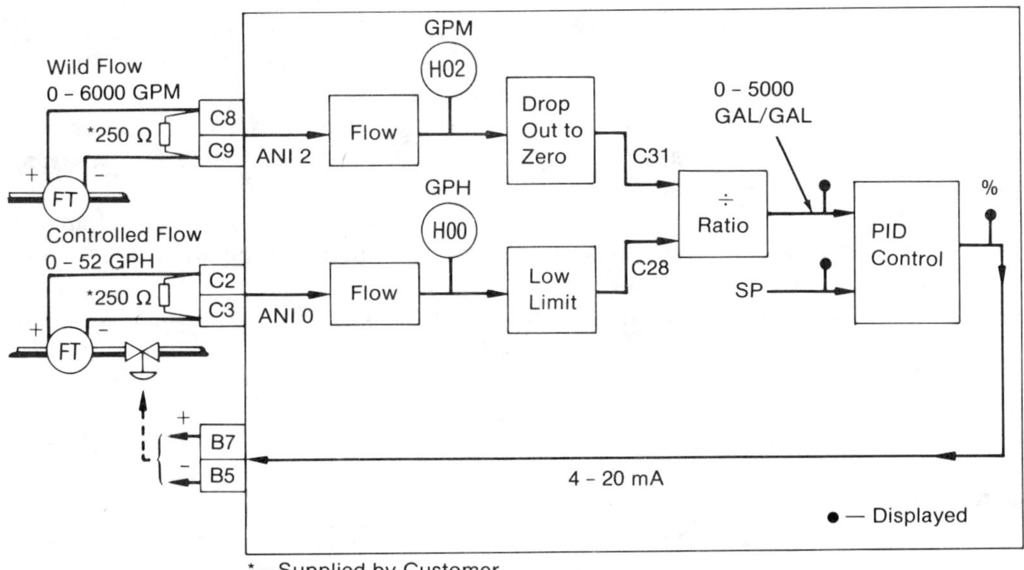

Figure 12-1. Ratio Control Loop Configured in a Single-Loop Controller

(Courtesy Fischer and Porter Company)

The following examples exemplify other types of "unmeasurable" values used in process control that have been developed using the Chameleon module. Each example can be created in a single instrument through programming.

1. Extensions to fuel/air ratio control, applied to combustion processes where fuel and air are ratioed. One option is biasing of the ratio value from percent oxygen measured in the exhaust gases to assure optimum excess air in spite of external load changes. Too little air causes smoking, too much wastes heat. The bias value is different for different firing rates, so its scaling should vary as the firing rate varies or as there is a change in furnace energy demand.

2. Cross limiting, another extension of ratio control. Cross limiting assures that air will increase before fuel when there is an increase in fuel required; but that fuel will decrease before air when there is a decrease in fuel called for. This prevents smoke in the exhaust gases but maintains the most economical mixture of fuel and air.

3. Mass flow. Gas flows are commonly measured as a function of differential pressure across an obstruction in the flow path. This is a volume flow measurement. The volume of a gas increases with temperature increase and decreases with pressure increase, so when these variables change significantly, volume flow measurements are not reliable. The weight of a standard cubic foot of gas, however, is always the same, no matter what its volume is. Mass flow, therefore, is a reliable parameter. A computation of mass flow can be made by taking the square root of the product of differential pressure and absolute pressure divided by absolute temperature.

4. Three-element feedwater control. Used for boiler drum level control, this control form combines measurements of drum level, steam flow, and feedwater flow to compensate for "shrink and swell," a false level indication introduced by steam pressure changes. It can also compensate for feedwater pressure changes.

5. Computation of the densities and enthalpies of saturated steam from pressure measurements. This can be done with third-order polynomials, constructed using the power, add, subtract, multiply, and divide operators. Temperature of saturated steam can be computed in a similar way from a pressure measurement and vice versa.

6. Override control. Sometimes two variables control different parameters of a process, and one or the other may be dominant. For example, flow from a tank will be controlled by throttling the rate of pumped outflow; but if an uncontrolled inflow threatens to cause an overflow, outflow must be increased on the basis of control to maintain tank level. Override control allows switching between these two demands bumplessly.

7. Computation of the contents of dished-head horizontal tanks from level measurement.

8. Batch control, on the basis of a sequence of timed and/or totalized flow measurements, interlocked with operating events.

9. A number of specialized proportional control implementations; these include gap with error squared, cascade control, feedforward control by set point computation, and feedforward control by output value biasing.

10. Multiple ranging of differential pressure transmitters for extending turndown of flow measuring range.

11. Automatic differential pressure transmitter calibration, to continuously update zero shift correction. This can also extend turndown capabilities.

12. Computation of supercompressibility of gases. This extends an orifice calculation from a point value to one that is useful over a large operating range of temperature and pressure.

13. Computation of supersaturation, used in the vacuum pan process for production of crystallized sugar.

14. Computation of boiling point rise, used in single and multiple effect evaporator systems.

15. Computation of completion of cooking in a pulp digester, by integrating relative cooking rate against time when upsets make the more usual temperature versus time control inadequate. In the pulp industry this is sometimes called H-factor control.

16. Generation of delay times to accommodate transport lag in control systems.

17. Sample and hold function for intermittent measurements. The Micro-DCI™ system includes single-loop controllers, Chameleon™ computers, and a Supervisor™, which is a processor and traffic director that allows modules to be combined on a highway and interfaced to a video monitor, programmable logic controllers, and/or a minicomputer link. The high level programming language can be applied to all three modules — 59 programming elements can be included in programming the controller, 400 elements in programming the Chameleon, and 1000 elements in programming the Supervisor.

Thirty-two modules can be included in a single network. The principle limitation is speed of transmission of information, which is limited because communication between modules is asynchronous, using ASCII characters.

12-4 Programmable Logic Controller Applications

Applications of programmable logic controllers are legion. Since PLC's take the place of relays they can be, and have been, used in every industry. Reports of applications can be categorized by function — material handling, machine control, etc.; by output type — turning devices on and off, servos, or variable analog signal control; or by industry. The functional category has been used here to present a few examples for each. The following applications are most common. There will be overlap in the example selection because many illustrations combine several functions.

1. Machine tool operation control, the original objective of PLC development. Applications include control of tool placement for boring, milling, drilling, grinding, welding, polishing, positioning, washing for chip removal, and gauging.

2. Material handling in many industries, such as mechanical assembly of automotive products and robotized operations, particularly for welding, parts placement and fastening, painting, tire balancing, and front end alignment. Supervising PLC's can download sequencing and timing programs depending on the amount and types of products, route material to its use point, and generate reports. Bar codes on parts and subassemblies are sometimes used to set up material requirements automatically. Separate PLC's supervise testing programs for the partially finished and final product assemblies.

Traffic control of accumulating conveyors in palletizing systems is another application of material handling.

Material handling includes conveying systems for handling of coal, ore, wood chips, grain elevators, and any bulk material that has to be accumulated in batches by starting and stopping conveyor systems.

Heat-treating processes, where material must be moved between furnaces and energy flow must be controlled on a time and temperature basis, constitute another example. The material handling part of a die casting operation might include automated part extraction, automatic die insertion, ladling of hot metal, and spraying of dies.

An interesting example of supervisory control of material handling is a car washing facility that measures the length of each vehicle as it enters, then adjusts the time for each step in the washing sequence to optimize the operation.

3. Sequential operations, a combination of time and event steps that may include starting and stopping equipment on the basis of interlocking, weighing, heating, cooling, combining, sampling, and counting. PID control is often used in these processes. Examples include:
- Regeneration of water softeners and carbon absorbers, back-washing of filters, rotation of cooling towers, blast furnace stoves, and precipitators.

- Batch chemical reactions, pulp digesters, coffee batch roasting and mixing, brewing, burner startup and management, and paint mixing and blending

4. Sequencing, combined with data acquisition at remote locations, with the information telemetered or transmitted through modems to a central processing area, falls into the category of supervisory control and data acquisition (SCADA). Applications include pipeline control, waste and water treatment, tank loading, gas distribution, pumping stations, environmental monitoring, and electric substation control.

12-5 Blast Furnace Stove Control

The applications listed above have been selected as representative of the discrete control aspect of PLC application. A list of process control applications equally varied could be made, but one described in some detail will serve to indicate the scope of programmable logic controllers for closed-loop control. Computational functions (add, subtract, multiply, and divide) in combination with the PID control algorithm are all that is required for this application, the heating of blast furnace stoves. This example was the first use of PLC's in the author's experience for a controlling activity normally accomplished with analog controls. A description of it will demonstrate the extent beyond on-off logical control to which PLC's can progress.

A blast furnace produces molten iron metal from iron ore, as heated air is blown through a column of ore, coke, and limestone. The combustion of coke, which takes place as the air passes through it, is incomplete. The gases coming from the top of the furnace contain carbon monoxide to the extent that, while the Btu content is low, blast furnace gas can be used as fuel.

Combustion of blast furnace gas is used to preheat air that will be blasted through the furnace for burning the coke. The air is heated in a stove — a tall vertical cylinder filled with brick checkerwork with a combustion zone along one side. Combustion of the blast furnace gas burned in this combustion chamber produces hot gases that pass through the checker work, transferring their heat to the bricks. When the bricks are uniformly hot and the heat has soaked completely through them, there will be a significant rise in stack temperature, indicating that the stove is ready to be used as an air heater. It is then closed up, or "bottled". Subsequently, blast air is blown through the hot checker work, cooling the stove as the heat is absorbed by the air. When it can no longer supply the needed heat, the stove is ready for the combustion (heating) cycle again.

There are at least three and sometimes four stoves used with a single blast furnace. While one is supplying heated air, another is bottled, ready to be used for heating. The other(s) are "on gas" — being heated by combustion of the blast furnace gases.

This heating portion of the cycle follows a specific pattern, as follows:

(1) Initially, gas is introduced at a rate controlled from flow measurement, and air is ratioed to it at an optimum fuel/air ratio.

(2) The dome of the furnace will reach a maximum safe temperature first, before the checkers have fully heated. Heating must continue but at a slower rate so that the dome is not damaged. To accomplish this the air valve opens wider and wider as the fuel valve position remains fixed so that combustion is continued with the fuel-to-air ratio continuously decreased.

(3) Sometimes the heating has not been completed by the time the air valve has reached its maximum opening. When this happens, combustion is continued with a continually increasing amount of excess air by slowly closing the gas valve as the air valve position remains constant.

During this entire period, the dome must be protected from excessively high temperature. Its measured temperature, through another controller, provides an override control that can take over from the combustion control if necessary. In this relatively straightforward control scheme, illustrated in Figure 12-2, there are three PID controllers for each stove; one controls gaseous fuel flow rate, one controls air flow at a ratio to suit the phase of the heating, and the third controls on the basis of stove dome temperature.

Combustion continues until the checkers can no longer absorb heat, as shown by the stack gas temperature measurement. When this temperature reaches a limit, the heating process is complete and the stove is bottled.

An extension of the control system may be used when the Btu value of the blast furnace gas varies or does not have enough heating value to bring the stoves up to temperature in the allowable time limit. This can happen, for example, when pelletized ore, which requires less coke per ton and therefore produces less combustibles, is charged. Pelletized or beneficiated ores have become common since World War II, when maximum capacity steel production exhausted the richest of the North American iron ore sources.

To get more heat for processing, the Btu value of the stove fuels can be boosted by mixing blast furnace gas with coke oven gas or natural gas. Control in this case is on the basis of a Btu/hour flow rather than a cubic feet/hour measurement. The value of Btu may be measured, or computed if measurement is not practical. Using this type of control increases the number of PID controllers for each stove to four, or more if several types of enriching fuel are mixed and stack temperature is also controlled on an override basis.

One more control loop must be included in the system. The stove, of course, will cool as air is blown through it. To maintain a constant hot blast temperature, the air to be heated is split between two controlled valves. One sends air through the stove being used for heating the air, the other bypasses cold air around the stove. The two streams, one hot and one cold, are mixed in proportions to provide a constant blast temperature.

Air to the furnace must never be shut off. For furnace safety, the valves are always manipulated so that one is always open. As the stove cools, the stove air valve will open; then

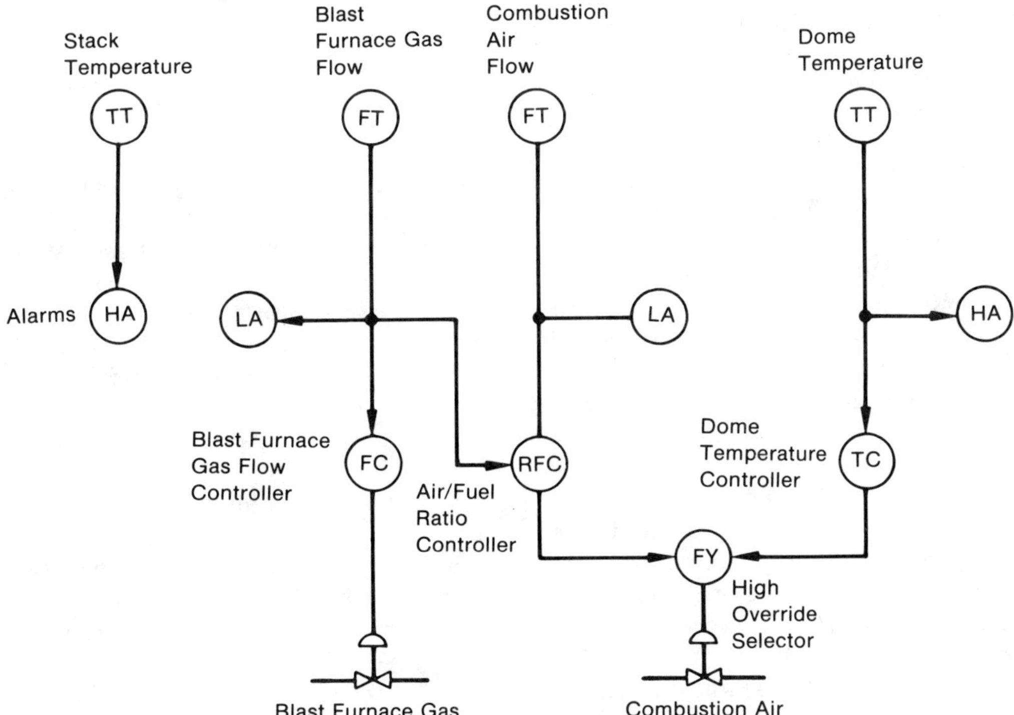

Figure 12-2. Basic Blast Furnace Stove Control Loops

the bypass valve will start to close. When it is nearly closed, the stove will have cooled to a point where it can no longer supply enough heat to maintain the blast temperature. At this point, it is taken off blast and a newly heated, bottled stove takes its place. The inlet and outlet valves of the cooled stove are repositioned for the heating cycle.

All of this valve switching (there are at least five large motor-operated valves for every stove), if done automatically, requires extensive logic systems for which a PLC can be programmed.

The installation witnessed by the author, where the analog controls had been replaced by PLC's, had three stoves. Only the stove heating, with fuel/air ratio adjustment, blast furnace gas flow control, and dome temperature override, was implemented. Stove heating was accomplished successfully, and the only criticism might be directed at the display of process information available to the operators. One LED display was shared by the three PID's for each stove, with indication of process variable, controller output, set point, gain and reset rates selected by push buttons associated with each controller. An operator is a very adaptive individual and can usually make anything work, if he wants to. In this case the operators took the requirement to select readings in stride, and operation of the stoves proceeded normally.

This installation was later replaced by a distributed control system used for a field trial. It, too, performed satisfactorily and had a much more informative display of information. It was not as dependable as the PLC, however, because of the environment in which the instruments were located. The control equipment was installed on the stove operating floor, a common occurrence in steel plants, exposed to heat, dirt, and rough handling. The PLC was much more ruggedly constructed than the distributed control equipment and was able to continue performing, while the latter could not until a special housing was provided for it.

12-6. Glass, and Glass Container Production Control

A description of instrumentation for glass making is included to illustrate distributed control system application.

Glass for containers is produced by melting a mixture of sand, limestone, chemicals that provide sodium, calcium, and other metallic elements, and cullet, the name given to recycled scrap glass. The finely ground mixture is heated, in a large brick-lined tub called a tank, furnace, or melter, by heat from flames produced by gas or oil combustion. At temperatures approaching 3000 degrees Fahrenheit the solid material fuses to form a viscous molten glass. The molten glass discharges from the melter through a basin called a refiner into a brick-lined canal called a forehearth. At the end of the canal gobs of molten glass are cut off and dropped into a container-forming machine, where bottles are produced. The bottles leave the forming machine on a conveyer that takes them through an oven where they are annealled to relieve stress and slowly cooled to a point where they can be inspected, counted, and packaged.

A typical combustion process produces heat to melt the glass by burning oil or gas with air. There is an optimum ratio of air to gas to produce heat without wasting fuel, at the same time producing no air pollution. After the gases have passed over the surface of the melter they still contain a lot of heat, and this is partially recovered by heating brick checkerwork underneath the melter. The checkers heat up, and the gases are exhausted through a stack to atmosphere.

Sets of checkers are located underneath the tank on the right and left sides. Similarly, there are burners for oil and/or gas on the left and the right sides of the melter. Firing alternates from one side to the other, and the hot gases are exhausted alternately through one set of checkers and then the other. Combustion air is brought into the burners through the set of checkers most recently heated, extracting heat from the checkers and cooling them. This saves some of the fuel required to heat the air to the temperature necesary for melting glass.

The level of molten glass in the melter is measured and then controlled by varying the speed of the mechanism that charges the mixture of sand and other solids into the melter. A measurement of the glass temperature at the refiner is used to control the amount of fuel being fired to a rate that will support the desired temperature. Air flow is ratioed to the fuel flow rate in order to maintain it at optimum rate for energy management, and the ratio is trimmed as a function of oxygen remaining in the exhaust gases. Both oil and gas can be used as fuels. A differential pressure measuring device is used to measure air flow, and since the air temperature varies as the checkerwork through which it passes cools, the measurement must be compensated to achieve a meaningful mass flow value.

Glass leaving the refiner must cool to a temperature appropriate to the size and wall thickness of the bottles being produced at any particular time. Temperature measurement is used to add heat from gas flames or cooling from air blown over the surface of the forehearth at several stations along the length of the canal. Since glass leaves from the bottom of the canal and heat is added or removed from the top, there is a complicated pattern of temperature levels in the glass in the canal. Finally, temperature of the air entering the annealing furnaces is controlled so that the proper amount of stress relieving is achieved during the bottles' passage through this part of the process. Production rates are determined by switches that send pulses to counters to show an accumulation of the number of bottles passing a point in the conveyer to the packaging area.

The majority of control systems for this process have been, until the late 1970's, made up of analog, panel-mounted instrumentation. As replacement parts for this instrumentation become unavailable, existing systems are being retrofitted to used new instruments, and single-loop digital control, distributed control, and computer-directed packages using pre-programmed minicomputers are being used. Programmable controllers can be used for the sequential operations required for furnace firing and exhaust gas reversal when all control operations must be suspended and a complicated sequence of valve and damper operations must take place. Many process variables in addition to those already mentioned, particularly of temperature, must be made, and so data acquisition is a system requirement.

Figure 12-3 illustrates part of an operational distributed control system used by a glass container-making facility. The section shown applies to the combustion portion of the control system, which uses gas and oil as fuels. The symbols identified as slots refer to time slots of 1/32 of a second, during which a processor will perform the function indicated. There are 16 time slots available in each processor in this installation. The processors are built on printed circuit cards mounted in relay rack assemblies in the melter area, and the displays and keyboards used by the operators are in a control room some distance away. Other processors in the same housing, as well as in other housings at other locations in the plant, collect data and provide control for furnace charging, forehearth temperature control, temperature control of the annealling furnaces, and bottle counting. If the distributed control system algorithm library includes a sequencer subroutine, as many of them do, furnace reversal can also be controlled using configured time slots.

Control in this illustration is accomplished by the movement of valves and dampers positioned by reversing drive motors (see Section 4-9). The output generated by each PID-configured time slot is a series of pulses that rotates a reversing motor clockwise or counter-clockwise to position a valve. A potentiometer, whose contact rotates as the final element is moved, acts as an internal positioning loop to tell the PID that the regulator has moved the amount (and direction) called for by the last control output. This type of control has some safety benefits, since if power fails the valves will remain in their last position. It is not necessarily typical of control in glass plants. Four to 20 milliampere outputs may also be used effectively for positioning air-operated valves equipped with current-to-pneumatic converters.

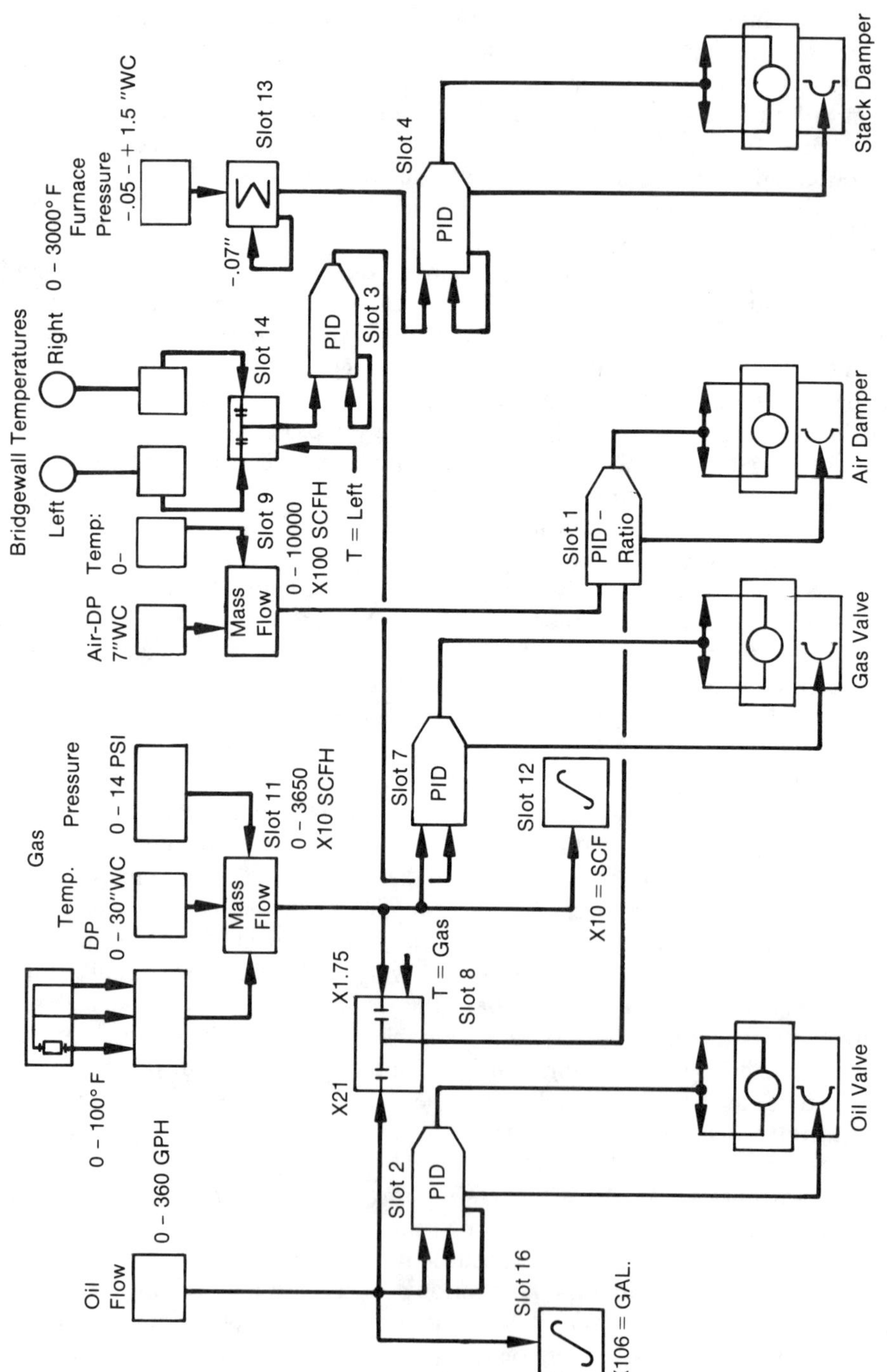

Figure 12-3. Distributed Control Configuration for Glass Tank Combustion Control System

In the portion of the plant shown, a gas flow measurement is computed by compensating the differential pressure across an orifice for temperature and pressure, configuring time slot 11 to perform the mass flow computation. When gas flow has been selected as the fuel, the compensated flow value is used to control gas flow rate by time slot number 7, configured as a PID algorithm that generates a control signal based on the deviation from a set point, which is a function of the deviation of bridgewall (at the entrance to the refiner) temperature from the desired refiner temperature. This control variable is computed in time slot 3. Different temperature measuring elements are used, depending on whether firing is from the left side or the right side, and the proper one is selected by time slot 14, configured as a selector switch. During left firing, a digital input to the processor is TRUE; this logical input is recognized by the time slot selector switch as a flag to select the input from the right bridgewall temperature measurement.

The amount of gas consumed per day is computed by time slot 12, configured as an integrator. The figure is used as a rough check on the gas company billings, as well as for furnace department accounting purposes.

When oil has been selected, its flow rate is controlled from the output of time slot 2. The quantity consumed is accumulated by integration of the flow rate using time slot 16. The pressure of the oil is held constant by control from another time slot, not shown, configured as a PID controller. Whichever fuel has been selected is used as the reference for ratioing air flow to fuel flow. When the selection is made, time slot 8, configured as a selector switch, gates the selected flow rate to the set point input of the air flow controller, time slot 1, configured as a PID-ratio controller. Since gas and oil require different amounts of air flow because they have different heating values, the selector switch algorithm scales the selected value to a Btu equivalent of mass flow rate. An air flow measurement, compensated for temperature to accommodate the changing checker temperature, provides the measurement that is compared to Btu input to achieve changes in air flow.

The ratio of Btu to air flow is a value entered into the configuration of the time slot 1. A measurement of oxygen in the exhaust gas, compared at another time slot configured as a PID, can generate a value that, as oxygen becomes higher or lower than a desired percentage, is used to bias the ratio value in the air flow controller to continuously change it to an optimum value.

Good combustion depends of the correct rate of removal of exhaust gases from the furnace. This in turns depends on maintaining an induced draft that will keep the atmospheric pressure above the glass slightly positive. This is the function of time slot 4, configured as a PID controller. The pressure cannot be measured at exactly the point in the furnace desired, in this installation, so a bias of the measured furnace pressure signal is created by time slot 13, configured as a $\pm$ summer, so that the operator will see a figure corresponding to the desired pressure, even though the measurement is somewhat in excess of this value.

Glass container plant control systems have a number of special considerations. Temperature control, particularly of forehearth temperatures, is extremely critical because of the relationship between production rates, bottle wall thicknesses, and precise forming requirements. There are many sizes and shapes of bottles in a plant's catalog of products, and each has its very unique temperature requirement for best production. Storing the control parameters as recipes for each type of bottle will assure repeatability and also reduce down time when the bottle-making machine is changed from one type to another.

Temperature sensing elements for measurements between 2000 and 3000 degrees Fahrenheit will be made using thermocouples or radiation pyrometers. Both of these devices produce signals that are not linear with temperature, and in the case of the radiation pyrometer may be affected by emissivity variations caused by smoke interference and flame variations. The nonlinearities can be taken care of by signal conditioning, available in the input circuitry of most distributed control systems. To completely get rid of emissivity variations may require special sensors. Digital control, with all of its benefits, will be only as good as the measurements it works with.

Different problems occur when thermocouples are used to measure glass temperatures. High temperatures will eventually cause deterioration of the couples, causing improper voltage outputs. A heavy protecting tube surrounding each couple will introduce lags in the response to temperature changes, and these must be recognized by adjusting the tuning constants of the PID controllers that regulate temperature.

Up to this point, the system described exists, implemented by distributed control. Many other uses for digital control are planned for this particular glass plant, making use of other digital control devices.

Some melting tanks use electrical heating, or at least an electrical boost circuit, in which an electrode protrudes directly into the molten glass. Because of the temperature of the electrode and the viscosity of the molten glass, cooling is required to prevent failure of the electrode. It is desirable to measure the temperature of the cooling medium for each electrode in order to detect coolant flow failure or leakage quickly before an element is lost. Monitoring the electrode voltage continuously is also advisable. A data acquisition system, in conjunction with a distributed control system with alarm displays, is one way of accomplishing this requirement.

Special problems exist in controlling forehearth temperatures to achieve the proper viscosity of the glob of glass dropping down to the bottle forming machine. Viscosity is a function of temperature and composition. The glob comes from the bottom of the forehearth channel. Regulation of temperature can change only the top layer of the glass, unless some form of mixing can stir the slow-moving, viscous material flowing through the channel. This is not easily done, and so serious lags will occur between the time of measurement and the time of correction, a difficult control situation at best. Ideally a model of the temperature pattern in the glass mass should be made to predict the required correction at one point to achieve desired bottom temperatures on the basis of measurements at other points in the glass. This can be approximated in a distributed control system, and accomplished more exactly by programming on a minicomputer that performs supervisory control through specific slots of a distributed control system connected to it over a data highway.

Combustion control on the basis of maintaining temperature does a very satisfactory job but cannot completely take into account all the variables involved. Variables such as heating value of the fuel used, ambient air temperatures and relative humidity, and the contribution to energy of the electric boost system contribute to the heat input to and loss from the furnace. It is recognized that better combustion control could result from continuous computation of a heat balance around the furnace, arriving at a Btu figure that could be used to control fuel flow, trimmed by temperature control. This would be a job for a supervisory minicomputer.

Another use for a minicomputer in conjunction with a glass plant system is the generation of periodic reports of process variables and operating parameters, printed out in a format to suit the needs of operation and management personnel. Many digital control devices can produce reports on a printer, but usually the format is preprogrammed and difficult to customize. Printouts per shift of electrode current, energy consumption, efficiency ratios, bottle count at various stations, production efficiency on the basis of globs of glass per finished number of bottles, fuel consumption, and any number of other values are headings for a customized report. The data for bottle count should by accumulated by a programmable control device because of the speed of counting required. A data acquisition system or the distributed control system can do much of the computation from raw data to obtain the ratios and efficiency values. Alternately, all of the computation can be done in the minicomputer program.

REFERENCES

1. Blickley, George J., "Smart Transmitters Monitor Tank Farm Inventories," *Control Engineering*, May, 1984, pp. 97–98.
2. Nixon, Steve W., "Recent Experience with Digital Process Control for a Glass Furnace and Forehearths," *Ceramic Bulletin*, Vol. 61, No. 12, 1982, pp. 1290–1291.
3. "Use of Digital Control in Glass Production," *GLASS*, May, 1983, pp. 171–172.
4. "New AGC Float Furnace Features Advanced Computer Control," *GLASS*, April, 1984, pp. 119–120.

UNIT 13
Process Applications — Systems of Devices

13-1 Digital Control Devices Used in Heat Treating Applications

A review of applications in a number of thermal process installations shows that users are beginning to apply the capabilities of digital control equipment. The following examples are installations currently in operation. Observe how these applications follow the pattern of need for recipe storage, report formatting, scheduling, and accurate control, attributes that form the basis for the excellence that allows digital control devices to perform so well in industrial process control.

Each of these installations has its own story. The first example is detailed; to save space and repetition, the others are summarized.

1. The application by a forging and heat treating facility of a programmable logic controller and single-loop digital controllers, each handling its share of activity but coordinated by a personal computer with standard and non-standard software, illustrates how the capabilities of a number of digital control devices can be combined advantageously to create an advanced system of heat treating controls.

This facility produces parts that must meet high quality assurance standards. After forging, parts are heat treated, quenched, and tempered to tight specifications for hardness, uniformity, and a consistent microstructure. Parts are specified to have a grain structure identical to that identified by tests from acceptable samples. For many products, for example, a Pitman arm (an automotive product, the connecting link between the steering column of an automobile and the tie rods that turn the wheels), assurance of quality is crucial.

To attain high quality, and at the same time to tighten control over production scheduling, the manufacturer decided to design a control system to satisfy his specific requirements. The installation consists of a single line — a hardening furnace for parts loaded by a vibratory feeder, a polymer or water quench system, and a continuous tempering furnace from which the heat treated parts are discharged into baskets. The two quench tanks are on a track, and the correct tank for the batch in process is moved into place as it is required. Furnace capacity allows up to 3000 pounds per hour of parts ranging from ½ pound to 10 pounds each.

The operational plant layout is illustrated in Figure 13-1. The three-zone furnaces are gas fired, with oil backup. Motor operated air valves are positioned from proportional controllers to maintain zone temperatures measured by thermocouples. Fuel follows air for ratio control. There is an overheat controller for each zone, monitoring a thermocouple separate from the control temperature couple. Each furnace, in addition to the three zones of temperature control, maintains furnace pressure at a few hundredths of an inch of water, positive so that cold air is not drawn into the furnace. This protects the work from oxidation, improves the furnace temperature uniformity, and reduces fuel consumption. Quench liquid temperature is maintained by pumping at a controlled flow rate through a heat exchanger, one in each quench tank. The cooling medium is water from a cooling tower, flow controlled to maintain a constant temperature of water returning from the exchangers to the tower.

The necessary control functions could have all been accomplished by programmable logic controllers. They also could have been done by distributed control. Neither choice would, however, have best satisfied an overall requirement for handling the many variations of heat treatment that correspond to many different products.

The control system design was based on a philosophy that if a part could be successfully heat treated, the same results could be achieved again by duplicating the treating conditions. Saving sets of parameters for many products requires a lot of memory, but with saved recipes

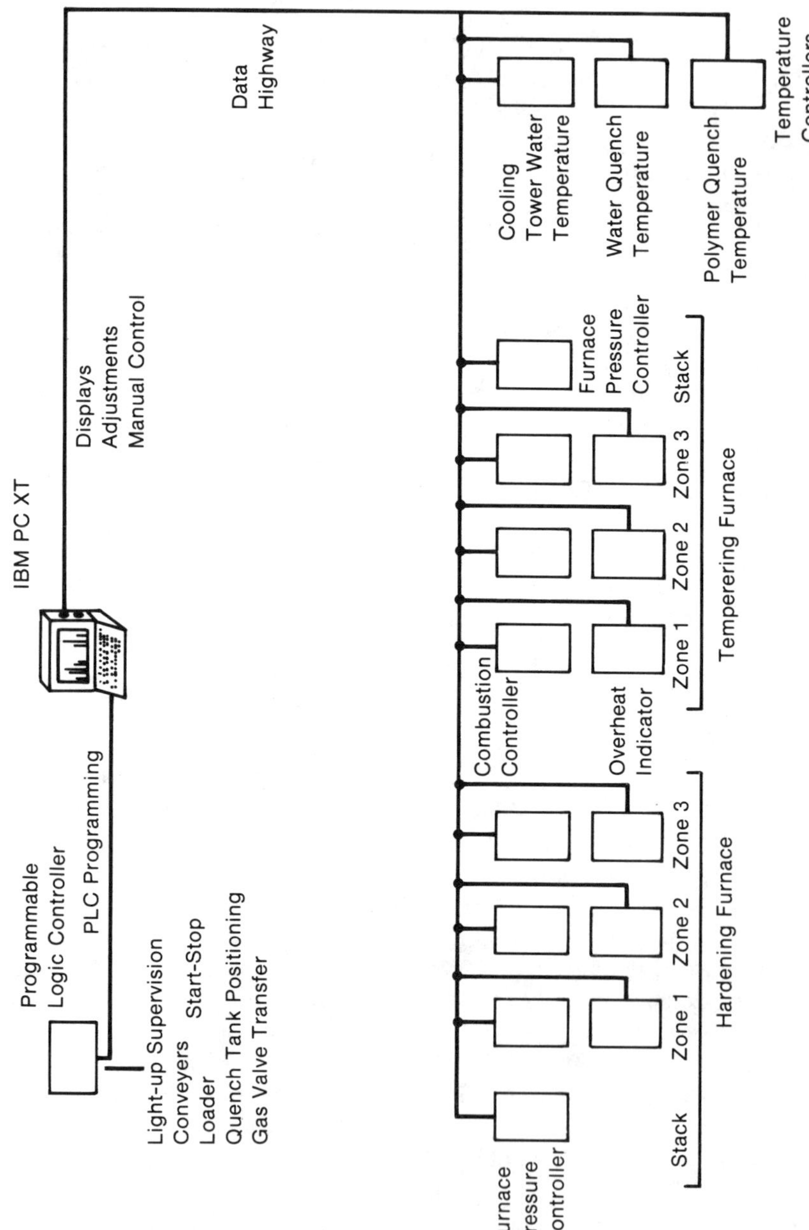

Figure 13-1. Heat Treating System

available the process for each specific product can be controlled so that it stays within the limits that yield a quality product. One reason for choosing the personal computer was to make use of hard disk memory capacity. The availability of the existing controller supplier's software programs for interfacing with single-loop controllers supported the choice.

The digital control devices are coordinated by the personal computer, an IBM PC XT™ with a hard disk drive, and a data highway. A programmable logic controller is used for startup and shutdown routines, and all material handling, quench tank movement, and furnace loading and unloading operations. The IBM PC is used to program the PLC; to store tables containing control parameters (set points, ramp rates, tuning constants, etc.) for heat treating many different categories of parts; to monitor the process and controllers and to display real-time data; and to prepare schedules for optimum material handling and energy saving. It will ultimately be used to analyze the data in the context of quality assurance and percent defects and to generate reports and certification for every lot of parts processed. A 30-channel strip chart recorder saves historical data, recording furnace and work temperatures for each furnace load.

The personal computer has been interfaced to 17 single-loop digital controllers. The controllers have capabilities that include tracking, remote mode changes, downloading recipes, and set point ramping, attributes desirable for this heat treating application and available only in a digitally implemented instrument. Also significant to this application, each controller has an RS 422/485 serial interface port, which can be used for sending real-time information to a computer but also allows information from the computer to be sent to the controller. Simply having the port means nothing — a program must be written to control the interchange of information. This program has been written by the supplier and is being used to provide an operator interface to all seventeen controllers in the system. The information passes between devices over a data highway. The control systems will function without the computer, which only looks at the current process status and provides operating guidelines. The controllers are independent and assure single-loop integrity.

Figure 13-2. IBM PC XT™ with Digital Single-Loop Controllers

(Courtesy Leeds and Northrup Company, a Unit of General Signal)

The digital control devices, PLC, single-loop digital controller, and personal computer are combined to do many tasks not normally possible for a heat treating facility. An example specific to the furnace operation is the accomplishment of large turndown ratios for optimum control of fuel-fired burner control systems at high operating temperatures (1600–1700 degrees Fahrenheit) as well as at low idling temperatures (200–500 degrees Fahrenheit).

Two gas valves, operating in sequence, are used to achieve the large turndown. The combustion air supply is controlled from 100% to a nominal 30% flow with a motor-operated butterfly valve. The gas supply is controlled from 100% to a nominal 30% with a zero governor regulator to give good control over the high temperature range. It is controlled from 30% to almost 0% with a separate motor-operated butterfly valve. This gives equally good control over the low temperature range. The problem is to switch from one to the other automatically, without causing a bump and upsetting the process.

During normal operation, from 100% to 30%, the temperature controller regulates the air supply, which is cross-connected with the zero governor control to regulate the gas supply. When air flow reaches its low limit (approximately 30%) a limit switch on the motor operator initiates a transfer through the PLC and the controller logic, switching the controller to the gas supply motor operated butterfly valve.

The limit switch action is timed in the PLC for five seconds, during which the instrument outputs signals to transfer the single-loop controller to manual mode, and to transfer the power and feedback slidewire connections from the air control valve to the gas control valve. During this period the control system is initialized by the automatic tracking feature of the digital PID controller, so there is no bump when the automatic mode is restored. After the five-second time period, The controller transfers to automatic mode, and temperature is controlled in an excess air condition, as the air is held at 30% while the gas flow is reduced to its low limit. During heat-up, the sequence is reversed, with the high limit switch of the gas butterfly valve drive motor initiating the transfer.

Monitoring operation is only the beginning of the usefulness of the computer. First of all, it has a hard disk with 10 megabytes of memory, an amount large enough to hold many operating programs, and many heat treating recipes. When a product has been processed through the furnaces and treated so that it leaves the tempering furnace with the desired hardness and microstructure, the controller parameters are downloaded to memory and saved. The next time the same product with the same specifications is required the identical control parameters that were successfully used for the original treatment are loaded into the controllers. The memory currently holds 50 or more recipes.

A second use of the personal computer is to run a scheduling program into which can be entered information about the day's or week's production backlog. There is no advantage in going from a material that requires low temperature hardening to one that hardens at high temperatures, then changing back to another lot that requires low temperature hardening. By grouping similar parts, the best efficiency is maintained. The program reduces lost time between batches to a minimum by matching finishing conditions for one batch with starting conditions for the next, within the constraints of delivery commitments. This allows a production schedule to be generated that satisfies delivery requirements, balancing the time to run against the differences of heat treating temperature profiles between adjacent batches. Energy consumption is reduced and production efficiency improved because time lost in bringing the furnace to a starting temperature is reduced.

The Lotus Symphony™ program is used to maintain a computer-stored notebook with information documented for every lot of parts treated. Each lot and record is sequentially numbered. The number of parts that went through, the weight, the time in each furnace, test results for the finished products, and records for any historical events can be recalled by date, lot number, heat code, or customer.

Other uses are being developed that will utilize the personal computer's full potential. The Symphony programming will be used to access files of process information from the PLC and the PID controller data bases, combining the information into new files that can be referenced from statistical quality assurance programs. These files will be used to generate certification documents and print out reports, automatically or on demand. The word processor capabilities of Symphony will be used to format the reports.

This application is an example of the potential for true system creation by combining individual digital control devices.

2. A carbon baking facility with twelve furnaces cycles heating and soak rates for many grades of product. Recipes are stored and downloaded as required, and formatted reports provide a record of every operation. As in the preceding example, a personal computer supervises single-loop controllers, twenty-four of them in this case, communicating with all units over a single data highway.

3. Recipe downloading is important for defining temperature and pressure relationships at different stages of production during the operation of a 300-ton hot press for forming high alloy material parts. In addition, inert gas pressures will be kept at values in relation to ram pressure that will optimize heat transfer and maintain uniform temperature throughout the work. A printer provides documentation for every piece formed. This is a minicomputer-supervised control system with single-loop controllers. It uses a multipoint recorder for recording and alarming, based on measurements of process temperatures and pressures.

4. Production through a carburizing facility, including a continuous carburizer furnace, a quench tank, a wash tank, and a draw furnace, is difficult to schedule because work differs in size, quantity, and specification. Because much of the control is a material handling concern, a programmable logic control is the master device. It regulates cycle times for the various sections of the system, changes recipes to suit material specifications, and changes set points for single-loop controllers that maintain temperature and gas flows.

5. A similar scheduling situation occurs for reheat facilities for steel and aluminum preceding rolling operations, where the mill products are constantly changing in accordance with customer order requirements. Different size and composition billets require different reheat cycles and close timing to have them ready for rolling, but not so early that energy must be used to maintain their condition. This is a very energy intensive operation; a mathematical model of the furnace continuously computes the thermal inertia of the furnace and of the work to optimize combustion and cause each piece of work to leave the furnace at conditions optimized for production and quality.

In this case, a DEC PDP-11™ minicomputer is interfaced to the steel mill's scheduling computer to coordinate the work flow. Single-loop digital controllers are used for three zones — temperature, fuel/air ratio, and furnace pressure are controlled.

The categories of digital equipment being used in these examples include data acquisition and loggers, programmable logic controllers, single-loop controllers, distributed controllers, minicomputers, and personal computers and multi-point recorders. The use of one or several of these categories is exemplified in each of the above examples. All have enhanced abilities for performing applications peculiar to the needs of this industry. The bottom line is reduced cost, increased productivity, and improved quality.

13-2 Pulp Mill and Paper Production Control

Pulp and paper mill installations were among the early industrial control applications of mainframe computers. When minicomputers became available at reasonable prices, many programs dedicated to control in specific process areas in pulp and paper production had been developed and are still being used successfully. It is estimated that today (1985) there are 3000 digital control systems in operation in pulp and paper mills throughout the world. In addition to computer-directed control, other digital control devices are used extensively. Programmable logic controllers have been used for digester control, distributed control systems are used for recovery boiler and lime kiln control, and computers are managing energy by coordinating turbogenerator-produced power and local utility purchases, monitoring demand control, and optimizing steam boiler and recovery boiler con tribution to plant steam load requirements.

A complete review of the instrumentation used for pulp and paper production would include discussions of pulp production, chemical recovery, energy management and recovery, environmental monitoring, and paper production. Review of the structure of a kraft pulp and paper plant, as well as of its control problems reveals needs that call for every benefit digital control can contribute.

In very general terms, pulp and paper production is accomplished by a number of separate but interdependent operations. These include wood chip preparation, digestion, using steam and chemicals, for making pulp from wood chips, pulp washing and bleaching, and the paper machine itself; and in the recovery loop, trains of multiple-effect evaporators, recovery boilers, causticization and lime recovery. Waste treatment and power generation are separate branches of the operation that must also be considered. The flow of production through these areas should interrelate so that each can operate economically and efficiently without having to wait for material from another area or having to exceed safe limits of capacity to make up for overload conditions. This is not easy because the demands for energy can be very random, so upset conditions are common. Between areas there must be compromise, otherwise each will have its own targets of performance. The washing area, for example, can produce the cleanest pulp by using excess water; but this throws extra load on the evaporation process, requiring, in turn, additional steam from the boilers. Ideally, a balance must be established. Control is very difficult in some areas, particularly in the operation of the digesters, the paper machine, and the recovery boilers, because special measurements are required and because there is much interaction of control requirements. Process modeling can improve controllability for interacting loops that is almost impossible to control automatically any other way.

Another problem making control difficult in pulp and paper making processes relates to process variable measurement. The rule that control can not be any better than the measurements on which it is based has already been mentioned. Many of the process variables are functions of material transfer rather than of energy transfer. Most conventional sensors are designed to work on the basis of energy transfer and are not satisfactory for measuring such variables as reaction completion, moisture content, or chemical composition. In the paper making process, for example, fast and accurate measurements are required for the weight of a cross section of the moving sheet of paper; for the moisture across the sheet at many stages of drying; for caliper, the profile of thickness of paper across the sheet; the hardness of the paper surface; opacity, color, etc. Special sensors have had to be developed to obtain continuous measurements of these key variables.

These production and control difficulties have been recognized as problems that require the sophistication of computer programming to create solutions that can be used effectively. Several suppliers have designed packages that include special sensors for the unique measurements. They have also developed mathematical models of the portions of the process where interacting loops and unusually long transfer times make control difficult. Many such systems are in use today for control of digesters and paper machines. In addition, systems of hierarchical control have been proposed for scheduling, energy management, production balancing, and optimizing. Only a mainframe computer with a coordinated program with access to all the real-time data in the plant can accomplish this sort of overall control.

The majority of the existing digital installations have been designed to solve the challenges of papermaking and of digester control, and they will serve as examples for this text. In addition (Reference 2), operation of recovery boilers and of causticizing areas, including the lime recovery kiln, are being optimized, using supervisory control distributed control at the operating level.

Paper Machine Control

At the beginning of the city-block-long colossus of machinery called the paper machine, a thin suspension of washed and bleached wood pulp in water is fed onto felt and then onto a travelling wire mesh through which the water drains to leave a thin, fragile sheet of deposited pulp. The sheet is many feet in width. It is picked up and supported by rollers, which pull it through the machine. During the passage, treatment produces a roll of paper that has a consistent dimension across its length, a hardness, moisture content, smoothness, and weight to satisfy production specifications. The paper moves at express train speed through the machine, and the consumption of energy is enormous. The flow of material is shown in Figure 13-3.

The rotation of large steel rollers moves the paper through the paper machine, coaters, winders, and calendars. The rollers are driven by electric motors, and there are many relationships between speed, current, and tension. There is a "draw" profile from one end of a machine to the other, and a speed profile also. The speeds of adjacent drives must be coordinated so that the paper is "drawn" from one location to the next. At various points in its passage the paper is under tension, controlled by having a section operate at slightly higher speed than the preceding section. Coordinated drives control the "nip" created between two rollers through which paper passes. Limits are critical. A seemingly empty building fills up with frantically busy people when there is a paper break.

Motor parameters can be controlled by SCR's with analog controller-developed inputs, or through the use of digital drive controls. Using digital motor drives permits master controllers to communicate with remote operator stations over data links to perform permissive logic for interlocking, multiple-drive sequencing, speed loop regulation, recipe downloading, and individual drive control.

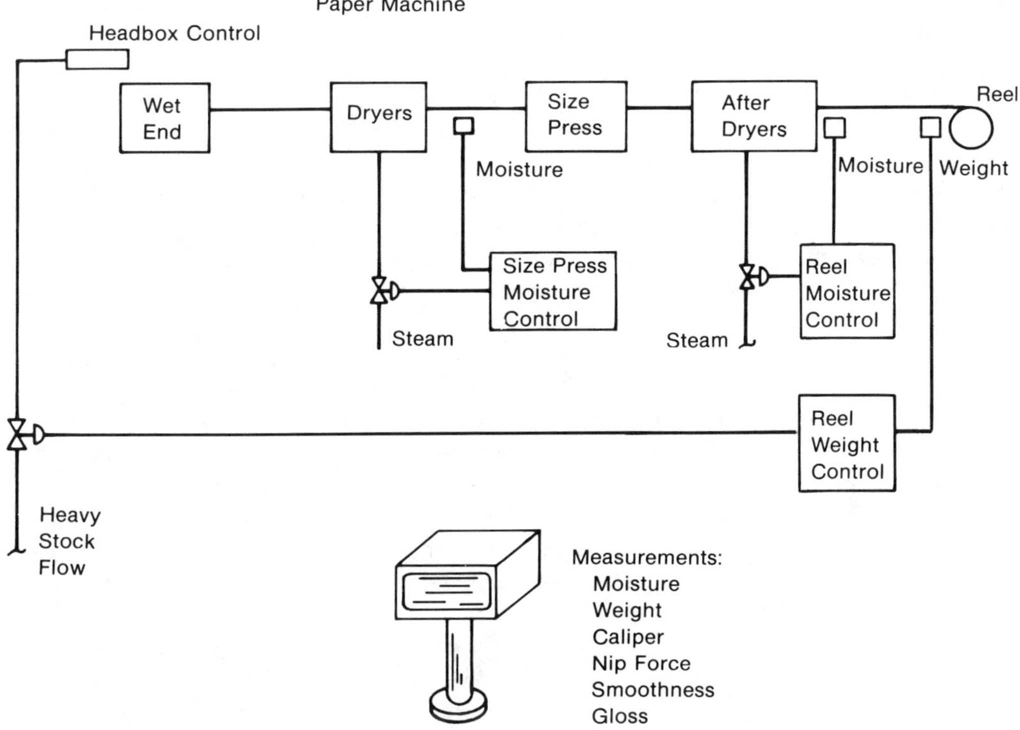

Figure 13-3. Paper Machine Controls

Digital control simplifies the coordination between motors to maintain speed and draw profiles for different grades of paper and stock material. When a recipe is changed the controller can recalculate draw values automatically, then change speeds in a section to match requirements in the rest of the machine.

There may be as many as 100 motors in a large paper machine. In addition to the need for coordinated control, large amounts of data handling for data acquisition, diagnostics, and alarms make a strong case for the benefits of communication links. Other benefits that can decrease machine downtime and increase productivity are storage and downloading of profile recipes, trending data to pinpoint conditions that cause paper breaks, and improved controllability through freedom from drift, with the flexibility to program branching to alternate control strategies when equipment malfuntions.

In addition to the operation of the machine itself, computer-directed systems for paper machines have three categories of control: weight, moisture, and caliper. Weight control is directed toward reduction of basis weight variations, moisture variation, and fiber. Control systems automatically level the distribution of paper stock (the thin pulp suspension) on the wire to optimize weight and moisture profiles. Measured values at each position can be compared to optimum profiles so that, in programming terms, arrays of set points and control actions take place simultaneously. Moisture control, particularly critical for the thicker paperboard and boxboard products, may require a balance between reducing speed or re-wetting the paper surface. Caliper measurements, using measurement heads that float above the paper sheet, are used to control caliper and smoothness profiles by positioning the space between rollers as well as the temperature of the rolls and their speed.

Digester Control

Digesters are used for chemical preparation of sulfate or kraft pulp from wood chips, the operation used in a majority of the mills in the United States. The sulfate process accounts for the characteristic odor that identifies the presence of a pulp mill in an area. In this process, wood chips are charged into a vessel called a digester, cooking liquor made of sodium sulfite and caustic soda is added, and live steam is introduced. After a period of time allowing for the hydrolysis of the lignins in the wood to alcohols and acids, as well as the physical separation of the wood fibers, the charge is dropped from the digester and sent to a washing process. The pulp is separated from the cooking liquor, then bleached and prepared for processing on the paper-making machine. The liquor is recycled through evaporators, concentrated to a degree where it can be burned to produce steam (in the recovery boiler), and the digesting chemicals are recovered.

The digesting process may be batch or continuous. Batteries of digesters may be used, introducing scheduling and inventory requirements that can be handled by a computer in the context of the requirements of other related department operations.

Figure 13-4 shows the principal parts of a digester control system. Control in the digestion process is based on the amount and composition of the caustic liquor, the temperature distribution throughout the digester, and the residence time. This is very long, and in terms of feedback control creates an almost impossible control situation. The ultimate target is the degree of delignification at the point where the pulped mass leaves the digester.

Important control loops used include wood-to-alkali ratio, liquor-to-wood ratio, chip level, production rate, temperature and wash flow rates. Every time a grade of pulp changes or a production rate is increased or decreased, many parameters in the control system must be changed to accommodate the new requirements.

Steam is a key raw material in a pulp and paper mill; it is used extensively to dry the paper being formed on the rolls of the paper machine, to evaporate the water introduced for bleaching and washing, and as the source of energy used to digest the pulp. Steam load is very erratic; paper machines operate continuously, but steam demand can change drastically when a machine is being started up or if there is a break in the paper. Many digesters are

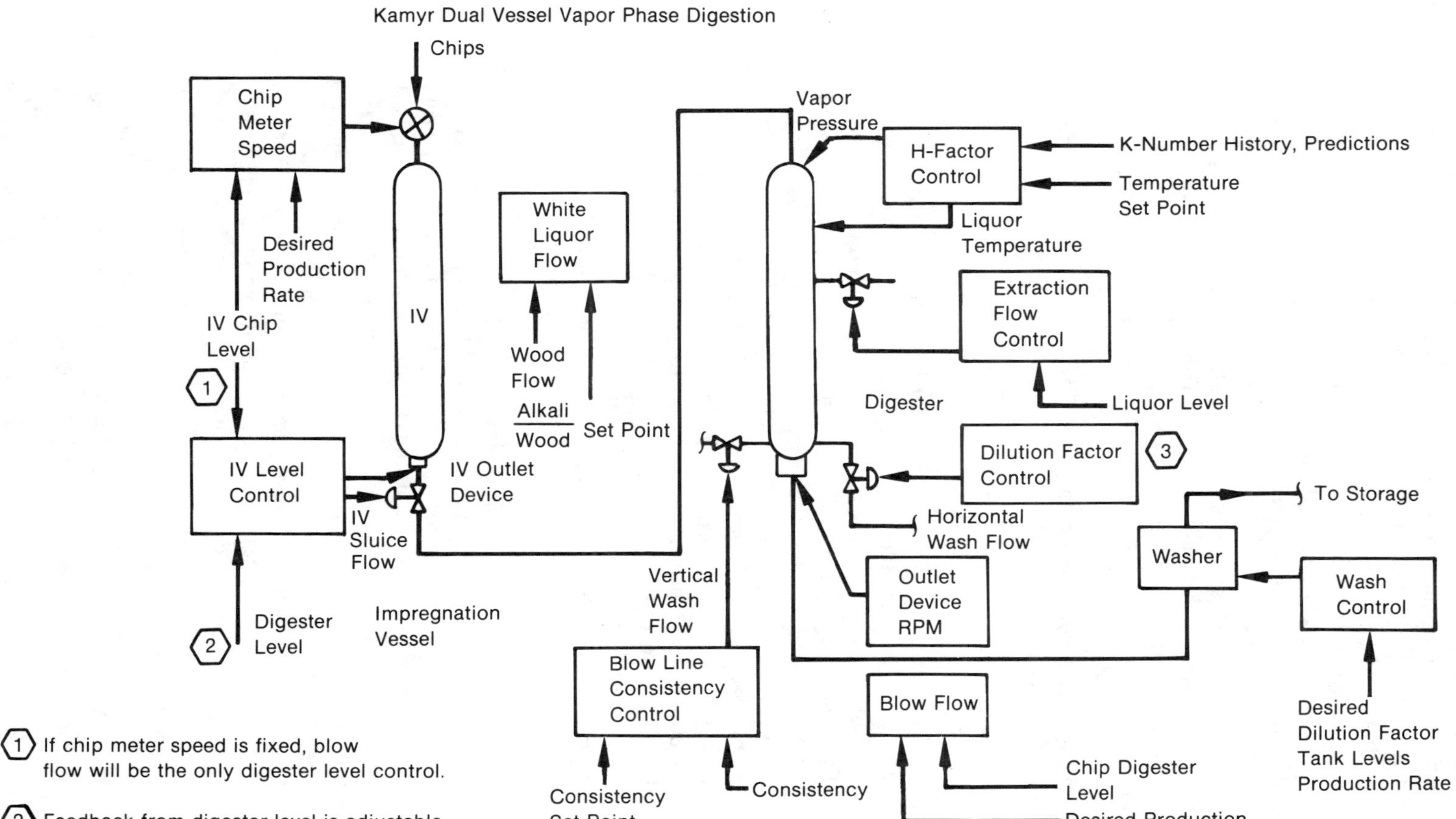

Figure 13-4. Pulp Digester Control Systems

operated on a batch basis; no steam is required during charging, a large amount is required during the heating period, then the rate is reduced during the balance of the digestion period. If there is a battery of digesters, the changing load may be evened out or it may be aggravated. In other words, the digesters contribute to the the inefficiencies of swinging boiler loads to meet operational demands. Scheduling of digester operations is another computer-directed activity that affects the productivity of the mill.

Digester control systems have been implemented using programmable controllers, distributed control systems, and minicomputers. Adaptive tuning has been used to take account of the change in controller response necessary as the conditions in the digester change. Because so many critical variables depend on computed values and on measurements of chemical composition for which there are very few reliable on-line sensors, and because of the tremendous time delays from the time of charging to the time for discharge, dynamic models of portions of the process have been developed, and their use has been proposed as an attractive solution. This makes the case for computer-directed control very strong. Gains to be realized from computer control are targeted toward higher yields, steam savings, decreased quality variations, reduced load on the recovery area, and higher production rates.

13-3 Waste Treatment

Like the pulp and paper industry, waste treatment has come to depend on digital computers for managing its control systems. Waste treatment sounds very simple; water is disinfected, solids are settled out, and the cleared effluent is pumped into a river. If it is filtered first, the filters have to be backwashed. Human waste is brought into contact with bacteria that reduce it to its elements, and the settling and filtering process is implemented. It seems as though the processes should go ahead under their own power.

The difficulties begin with the input flows. There may be four different kinds to deal with: human waste, industrial waste, residential and industrial water, and ground water. The inflow is random; an unexpected or prolonged rainfall can completely upset the balance of flow. Immediately, a problem of inventory must be met. There must be available storage capacity to absorb increases and allow reasonably constant flow conditions through a plant that handles larger volumes of processing material than any industrial plant could sustain. The methods of treatment must also change, to accommodate the new mixture of influents. At the same time, processing is slow, and long delays for conversion and settling reduce certain portions of the process to a self-controlling status. The inventory problem is compounded by the economics of power usage, demanding that large horsepower motors be turned on selectively so that they are not used during peak power load periods. A few minutes of pumping at rates of tens of thousands of gallons a minute during a high rate period can set a level of payment for power for the entire payment period that is devastating to operating costs. Normal pumping patterns change constantly during a 24-hour period, as there is an ebb and flow from one source of contribution, then from another.

For modeling the process created by this inventory change alone, use of a computer is recommended. Added to this are the requirements for copious reports and records, necessary to demonstrate observance of local and federal government regulations. This demands the collection of many process variable values, constant updates, processing of the data acquired, and formatting of the reports.

Quality control must be rigidly observed, and the specifications to which product must be maintained are tighter than most industrial establishments need to work within. Where tolerances of $\frac{1}{4}$ or $\frac{1}{2}\%$ may be difficult to maintain for industrial grade products, effluent contaminants are allowable on the basis of a very few parts per million. Reagents must be added within carefully controlled limits; costs are high for reagents necessary to treat many industrially produced heavy metal wastes; but more importantly, too much of some reagents can be as ineffective as too little. Aeration is important for protection of bacteria colonies.

Failure to maintain correct limits of oxygen environment can destroy bacteria, which cannot merely be replaced — replacements must be grown. The solid products of the waste treatment (sludge, in particular) must be disposed of. This can introduce additional problems associated with incineration, a complete system in itself with its control and scheduling requirements.

Finally, centralization of operation and control is desirable because of the area covered by a waste treatment facility. Coordinating information from a number of remotely located pumping collection stations, clarifiers, banks of filters, and all the other treatment areas would be difficult and expensive in manpower, if it had to be done manually.

Digital communication and processing of information is a logical way to handle these problems. All of the categories of digital control devices that have been discussed in the text can be combined over data highways: minicomputers for the sophisticated modeling and report generation, data acquisition systems for the many data entries, distributed control for the operator interface and control, and programmable logic control for sequencing, as required for filter back-washing.

A brief summary of the instrumentation of a typical industrial waste water treatment facility is included in Table 13-1, to put the variety and type of instrumentation into context. The areas of treatment are shown on Figure 13-5.

It is not the intent of this example to recommend any system for performing these actions, but simply to show the mixture of requirements. It does seem reasonable to suggest that use of minicomputers — programmed to take into account the problems of allocating flows, holdup times, and equipment usage to accommodate the widely changing load conditions under which a waste treatment facility must operate — offers the best way to achieve optimum efficiency and profitable operation. The computer's ability to format reports can also be useful. It is the intent to point out that whatever type of instruments are selected must fit into a system, even a hierarchy of control and measurement, that can be implemented by some combination of digital control devices and networks.

13-4 Energy Management Control Systems

The use and production of energy affects the costs of all industry. It is a prime example of the way combinations of interacting variables and process load changes, in conjunction with economic considerations, create a control situation that is simply beyond the capabilities of standard analog control equipment to handle automatically. Digital control devices can fill this need by performing the control and making the computations and decisions in many industrial installations.

Before the oil embargo in 1973, not much attention was paid to reducing energy consumption and saving fuel. Much more concern has been given to the matter of energy conservation in the 10 years following the fuel shortages. Saving energy has proven to be a high payback opportunity for updating control systems and applying sophisticated control schemes. Significant reduction in oil imports is evidence of the success of this endeavor. It can be assumed that savings to be gained by managing energy usage will still be sufficient to promote a continuation of the development of control applications for saving energy.

All categories of digital control devices are contributing to the activity. Applications involving individual machines or processes are being satisfied by data loggers and single-loop controllers. Larger complexes of processes are being handled by distributed control systems, connected to central control areas by data highways. PLC's have a place also, primarily in local analog control applications where they are competitive with distributed control or can share a data highway but also for material handling and logical control applications. The biggest savings, because of the complex nature of the problems, come about by optimizing and allocating resources, and the larger minicomputers are the only digital devices with the memory size and speed to perform the computations that must be made to be successful.

Table 13-1
Control Functions Associated with a Waste Treatment Facility

Process Function	Process Variable	Readout			Control		Alarm			Logic	
		Record	Indicate	Integrate	Proportional	Digital	High	Low	Malfunction	Sequential	Interlock
Raw water											
Mechanical screens	dP		x			x	x				
Low service sumps	level		x			x	x				
Low service sumps (3)	status					x					
Raw water	temp.		x								x
Mechanical screen	status					x			x	x	
Storage	level	x	x				x	x			
Raw water pumps (3)	status					x					x
Raw water pumps (3)	pressure		x								
Raw water	flow	x	x	x	x						
Activated carbon	addition		x		x						
Flocculation and clarification											
Rapid mix											
Chemical feed pumps (3)			x		x	x					
Clarifier											
Mechanical devices			x			x					
Prechlorination	flow		x		x						
Recarbonation	pH		x		x		x	x			
Filters											
Influent	level	x	x		x		x	x			
Loss of head	dP							x			
Rate of flow	flow	x	x		x						
Backwash						x				x	x
Clearwell	level	x	x				x	x			
Effluent turbidity			x				x				
Chemical feed											
Residuals	flow	x			x		x	x			
Fluoride feed	flow		x		x		x	x			
	conc.	x	x				x	x			
High service water											
Storage tank	level	x			x		x	x			
High service pumps	pressure	x			x						
Discharge	flow	x		x							

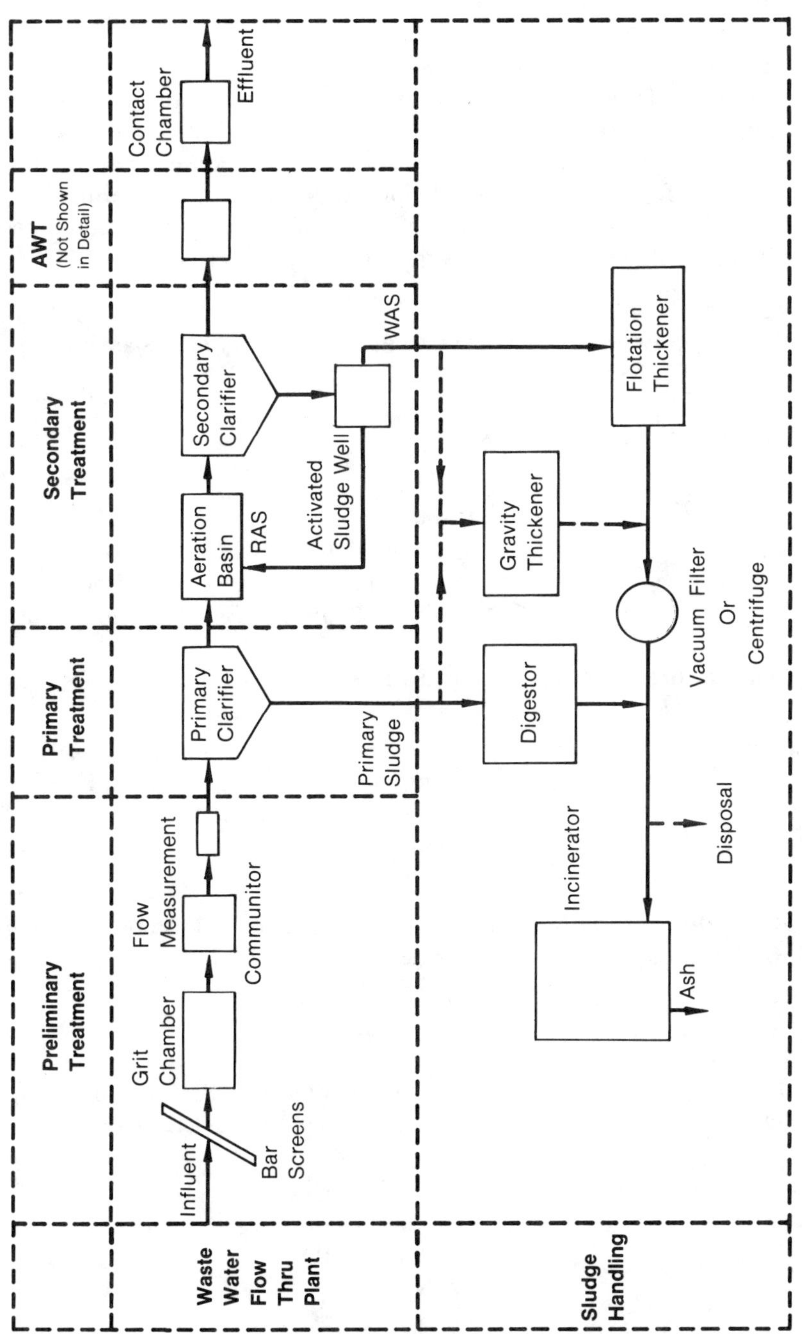

Figure 13-5. Conventional Municipal Waste Treatment Plant

(Courtesy Fischer and Porter Company)

This section presents an overview of the energy management opportunities that exist in industry. The nature of the process requirements and the many interacting influences should be an indication, even with a very condensed presentation, that this is an area where digital control must be used to achieve acceptable results.

Table 13-2 lists the various applications associated with energy management in relation to the digital control devices that have been of use in effecting savings. This is not to imply that the categories listed are the only way or the best way to accomplish control. A review of the literature written during the decade when energy management activity was highest indicates that these methods have been used. Personal computers are not included in the tabulation, partly because there is no history of successful application, partly because the most attractive computer applications are for plant-wide installations and require extensive memory and programming. This does not mean that optimizing for an individual piece of equipment or process could not be handled adequately from a personal computer.

Table 13-2
Power Generation Activities Controlled to Support Energy Management

Process Function	Applicable Digital Control Equipment
Power Generation	
Steam boilers	
Control	Control of header pressure, combustion,
Header pressure	drum level, 02%, blowdown, and heater
Drum level	outlet temperature have been done with
02%	PID, PLC, and DCS. 02% and combustion
Blowdown	control with SPEC. Burner management
Heater outlet temperature	with PLC's.
Optimizing	
Steam generation	Optimizing and allocation programs have
Fuel allocation	been written for minicomputers, based
Load allocation	on data collected by DAQ.
Electric Power	
Optimizing	
Turbo generator load allocation	The same is true for electric power and
Extraction flow allocation	for tie line control.
Cogeneration	
Power usage	
Tie line control	
Demand control	Demand control and load shedding has
Load Shedding	been done by PLC's, DCS, MINI,
	and SPEC.
Other Processes	
Cooling towers	Fan sequencing — PLC's and DCS
	pH loops, PID
Refrigeration	MINI
HEVAC	MINI
Incinerators	PID, PLC

Key:			
DAQ	Data Logger	DCS	Distributed Control
PID	Single-Loop Control	MINI	Minicomputer
PLC	Programmable Logic Control	SPEC	Special-Purpose Computer

The functions listed in the table are briefly described below.

Steam Power Generation — Steam Boiler Control

Header Pressure Control

Boilers generate steam by boiling water at a rate and pressure that satisfy the steam user's load requirement. Measurement of the pressure in the header into which steam produced by one or more boilers discharges is the basis for controlling the steam production rate. The efficiencies of the boilers producing the steam is a determining factor in setting the firing rate for each of the boilers. Common practice is to fire the most efficient boiler continuously (base loading) and "swing" other boilers in and out, following the demand as required. An allocation program prevents boilers from being overfired or underfired by gradually shifting the load on the basis of economical optimization calculations. In addition to assuring better efficiencies, there is less damage to equipment because of reduction in overheating of tube surfaces and overcooling of combustion zones.

Combustion Control

Based on the header pressure, fuel is burned to produce heat, which then boils water in a boiler drum. There are well-established strategies for combustion control. Not only must the header pressure be maintained, but fuel and air must be mixed in a ratio so that enough air is provided for combustion of the fuel, without producing smoke, but not so much that fuel is wasted in heating it.

In most systems, fuel flow will be regulated on the basis of header pressure measurement. Air flow will be ratioed to fuel flow, following it. (Some systems control air flow, and ratio fuel flow to it). Furnace pressure is usually controlled as a separate loop to maintain best draft conditions for good combustion. There are variations:

(1) Cross limiting. When load increases, increase the air flow before the fuel flow increases; but when load decreases, decrease the fuel flow before the air flow decreases. This assures smoke-free firing for a small increase in excess air.

(2) Feedforward control of combustion and of furnace pressure, based on boiler load changes. This is done so that when changes are large and sudden, some anticipatory action can be taken.

$O_2\%$ Control

Excess air must be used to assure complete combustion of fuel, but for most efficient combustion the excess should be as small as possible. The amount is not a fixed percentage; it varies with firing rate and the type of fuel. Conditions external to the combustion process (combustion air temperature and humidity, for example) have an influence on the amount. It is advantageous to change the fuel/air ratio as needed to maintain optimum excess air. This can be done by measuring percentage of oxygen in the exhaust gases, comparing the measurement to a set point value and using the deviation from set point to bias the fuel/air ratio setting used in the combustion control system. The oxygen percent set point is changed, in turn, as the firing rate changes, so that the optimum value is always called for.

Heater Inlet Temperature

Exhaust gases leaving the boiler through the stack have been heated by the combustion of the fuel, and this heat is lost. Some may be recovered by using it in a heat exchanger to heat incoming combustion air. If too much heat is removed, the stack gas temperature can drop low enough so that moisture will begin to condense. Condensation, in conjunction with sulfur and nitrogen elements in the gas, creates a corrosive condition harmful to the stack, so there is a definite limit to the amount of heat recovery.

Blowdown Control

Feedwater to the boiler drum leaves the system as steam, but solids used to treat the water remain behind, and their concentration in the drum accumulates. There must be a periodic removal of solids to protect blades of turbines driven by the steam (carryover of solid material can destroy the blades) and to prevent the boiler drum from eventually filling with solids. This is accomplished by venting water under pressure, blowing it and its accompanying solids down the drain. Heat already used to raise the temperature of the water to the boiling point is lost, so there is an economy in regulating the amount of blowdown in accordance with a measurement of conductivity.

Boiler Drum Level

Feedwater is added to the boiler drum on the basis of a drum level measurement to make up for the steam produced and removed. There can be considerable interaction between level, steam flow, steam pressure, and feedwater pressure, depending on the size of the drum; the variations in steam demand and feedwater pressure; and the rate of steam production. Large capacity, high pressure systems use a cascade control, with the level controller output signal becoming the set point to a feedwater flow control loop. At the same time steam flow introduces a feedforward action that compensates for lag in the response of drum level to load changes.

Flame Safety and Burner Control

Gas-burning boilers use a complicated system of flame detection devices, working in sequence with relays and timers to assure that combustion of gas is initiated and maintained in accordance with the regulations of boiler safety codes.

Optimizing

There are many opportunities for effecting significant saving by optimizing in steam producing systems that have multiple fuels, multiple boilers, or multiple loads.

Many steam power installations have several kinds of boilers that can be fired with a variety of fuels. Gas — natural, coke oven, and blast furnace; oil — heavy and light; coal; and bark are fuels used for steam production. All have different heating values, and their relative availabilities, combined with the varying efficiencies of the boilers that use them, require a very complicated computation to find the most economical combination. Kraft pulp mills have the added complication that a large percentage of the steam they use is produced by burning the combustible chemicals from the wood pulp digested to obtain the cellulose from which paper is made. At the same time that steam is produced in these "recovery" boilers, chemicals used in the digesting process are also being recovered.

Other Sources of Heat

Incinerators, where heat is generated from the burning of trash, chemical plants, where heat of reaction sometimes produces enough energy to make steam, and jet engine exhaust gases used in waste heat boilers provide alternate sources of energy. Each has its unique control requirements. The contributions of each of these sources are factors in optimizing the production of steam.

Electric Power Generation

Steam produced at high pressure can be expanded through reducing valves to meet lower pressure requirements. Alternatively, it can be expanded through turbines that generate electric power, and lower pressure steam is produced at the same time. Some plants produce as much electric power as they can in this way for their operation and purchase the balance from a local utility through a tie line. There are many energy saving opportunities through turbogenerator network management. Considerations that must be satisfied are whether to make or buy power and how to allocate the kilowatt load to different turbogenerators as steam loads and electrical loads increase and decrease, on the basis of the economics of condensing and extracting steam at various pressures. This can only be solved by extensive optimization, based on calculating heat rates, mechanical efficiencies, and steam path efficiencies. A good solution requires understanding of the process requirements, the plant economics, and the characteristics of the equipment.

Power Demand Control

Purchased power costs are based on usage over a period of time. The utility selling the power monitors the usage. If a limit is exceeded, the cost of all power purchased increases. One common method of monitoring is to read peak wattage values at one-minute intervals and average them for a 15- or 30-minute period to compute the average peak demand for that period. When a new period starts every minute, this is called a "moving window". The periodic computation is a basis for billing, and the demand charge for electricity is based on the highest kilowatt hours used in any demand interval. The top rate for a short period can set the billing rate for a period of one or even several months.

Demand control systems, designed to meet the characteristics of the billing structure for the line they are used on, can have a considerable impact on power costs. Following the trauma of the oil embargo, creation of small, special-purpose computing devices proliferated. These were designed to monitor power usage and control its use by dropping non-essential loads or by reducing consumption. The special-purpose designs were difficult to standardize at economical prices because each plant requirement had to be tailored to its process requirements, to the availability and type of load that could be taken out of service, to the contract with the utility, and to the availability of measurements for acquiring power consumption figures. Installed systems using more standard digital control devices ranged from data loggers (Reference 5) with computational power surveying plant conditions on a continuous basis to provide historical data so that projections could be made to control peak demands, through PLC's, to minicomputers.

One method of power demand control computes an ideal rate for the period, then matches the actual accumulated rate. If the ideal rate is going to be exceeded in a window period, the demand will be reduced by shedding a succession of electrical loads, if production requirements can be sacrificed. This is feasible if a number of large loads (electric furnaces or large horsepower motors, for example) are part of the process equipment and can be turned on and off without causing harm to the product.

Another form of power demand control puts on all loads at the beginning of the demand interval and starts to accumulate actual consumption values. At small intervals, the system predicts whether or not the pre-established limit will be exceeded if the current instantaneous rate is continued. It sheds and restores loads, depending on whether its calculation shows that too much or too little power is being used.

Still another method, independent of the actual demand interval, measures the rate of consumption at frequent intervals and compares this rate with a preset limit. Loads will be shed at any time the instantaneous rate exceeds the preset rate. The method is more expensive because it does not try to squeeze as close to the allowable limit as the other methods, but it is easier on equipment because it maintains a more even rate of equipment operation.

Other Processes

Cooling towers, refrigeration systems, and building heating and air conditioning systems are all prospects for optimizing and control. Thermodynamics, heat transfer, flow allocation, and load allocation are factors in all of these systems; therefore, there are computed variables that can be significant and calculations that can be used to determine the optimum combinations of operations. While the size and control complexity is not equivalent to that of a steam power or electric power system, it can be extensive enough to require the services of a large computer to compute optimizing routines and supervise the control operations in order to get benefits from energy management.

Documentation

Digital communication and monitoring provides opportunities for saving costs as a corollary to the information that can be collected and formatted for energy accounting reports. Careful study of the information collected will often reveal efficiencies that are apparent when energy balance figures are examined. If the information is not collected and examined, energy-wasting practices may go unobserved for indefinite periods.

13-5 Hierarchical Control

Hierarchies of tasks to achieve complete plant automation have been mentioned several times in this text in connection with applications for mainframe and minicomputers. This extention of distributed control is not yet a fulfilled idea, but it promises so much benefit that it must soon become a reality and deserves to be reviewed as a bona fide application. Its aspects have been well documented in Reference 11, and much of the following description is abstracted from that reference.

Distributed control (and this time, the phrase must include programmable logic control) accomplishes tasks on three levels of activity.

The first level — specialized controller modules, directly interfaced to the process, measure and control.

The second level — a central operation coordinates the duties and activities of the first level units.

The third level — minicomputers in a supervisory capacity enhance first level operation by the application of sophisticated programming. The activities of the second level can also be monitored and directed in the event of actual or potential emergencies.

Levels 1 and 2 together, or 1, 2, and 3 together, can function on a stand-alone basis.

The hierarchy can be extended to higher levels to achieve the goals of maximizing plant productivity and minimizing costs and energy usage within the framework of scheduling production, marketing products, purchasing, and personnel assignments. The overall objective is to assure reliable production and deliveries of products when they were promised to the customer, while at the same time supporting management objectives and decisions.

Level 4, then, would oversee a number of associated level 1, 2, and 3 combinations to coordinate overall production schedules, maintain inventories to satisfy foreseeable requirements, and allocate raw materials, energy, and manpower in the context of departmental operation.

The activities in a steel mill provide an illustration. Iron is produced in a blast furnace, refined to steel in a basic oxygen furnace, then cast into ingots. The steel solidifies but must be reheated, brought to temperatures suitable for rolling into plate, strip, and wire, and soaked at the desired temperature until the mill is ready for them. The rolled products may be annealed, galvanized, or plated.

Each of these many operations can have its own level 1, 2, and 3 control systems. There must be many compositions and sizes of billets to satisfy a multitude of customer requirements; great expenditures of energy to prepare the steel for processing and to hold it in a condition that allows it to be worked; and much scheduling to balance the need to keep the rolling mill in operation without having to maintain a large reserve of heated steel ahead of it. To manage this complicated mix of variables effectively requires real-time knowledge of process conditions in each operating area as well as reference criteria from overall plant management.

A large plant complex will have a number of departments, any or all of which could have the complexity of the steel plant example. The uppermost level in the hierarchy, level 5, coordinates all of these in the context of management objectives. Programming has to accommodate inputs from marketing, purchasing, finance and accounting, and production support units (power, waste treatment, engineering and construction). Part of this input may come from an interface to a management resources planning program, although this is usually a static study and not based on real time.

There are a number of things happening that indicate that hierarchical control may soon become a reality. While Reference 1 is certainly a most serious proponent of the concept, it is not the only one. Several instrument companies are actively promoting the idea. Honeywell is presenting its TDC-3000™ distributed control system design as an ingredient of hierarchies, combining the expertise of its business marketing division with that of its process control activity; Allen-Bradley is advertising the Productivity Pyramid™ concept as a systems approach to automation; Foxboro and Hewlett-Packard have implemented an agreement that will combine the former company's process control experience with the latter's strength in computation and data analysis instrumentation; almost every distributed control system supplier has an option for a computer interface link, which opens the door to the establishment of a hierarchy of control.

The Allen-Bradley design is described in company literature in terms that indicate that programming has been done for a complete system. A-B literature describes "a hierarchical system for control, monitoring, on-line testing, and scheduling of an automated assembly manufacturing cell." Five levels of control are defined.

Machine Level	Sensors	— Photosensors, proximity switches, limit switches, etc.
	Actuators	— Motor starters, solenoids, dc drives, etc.
	Identifiers	— Bar code readers, badge readers, etc.
	Ranges	— Vision systems, pressure transducers, etc.
Station Level	PLC-2/30™ PLC controller;	
	Machine controls are configured with a standard data table for alarms, production information, reasons for product rejection, service response, parameters, performance information, and system control and status.	
Cell Control	PLC-3™ Master PLC controller, which collects data from the station's controllers and performs data compression with the central control computer and dispatches production orders; Color graphics; Maintenance and operator terminals.	
Center Control	VAX 780™ computer, which collects and stores data from the cell level controllers, annunciates alarms, provides production reports, promotes interdepartmental communications, and calculates statistical quality assurance data; Operation Terminals.	
Plant Level	IBM mainframe computer prepares production and status reports on the basis of production order data from the central control computer; Printers.	

The information flow is illustrated in Figure 13-6.

This is not included as any kind of endorsement of a product, but as an example of the organization that must be undertaken to make up a hierarchy of control. It supports the indication that programming for this application is beginning to appear.

The technology to support hardware and software design certainly exists. It is not clear who is going to supply software, however. The objection may be made that each facility that wants to install a hierarchy of control will have its own unique problems, structure, and

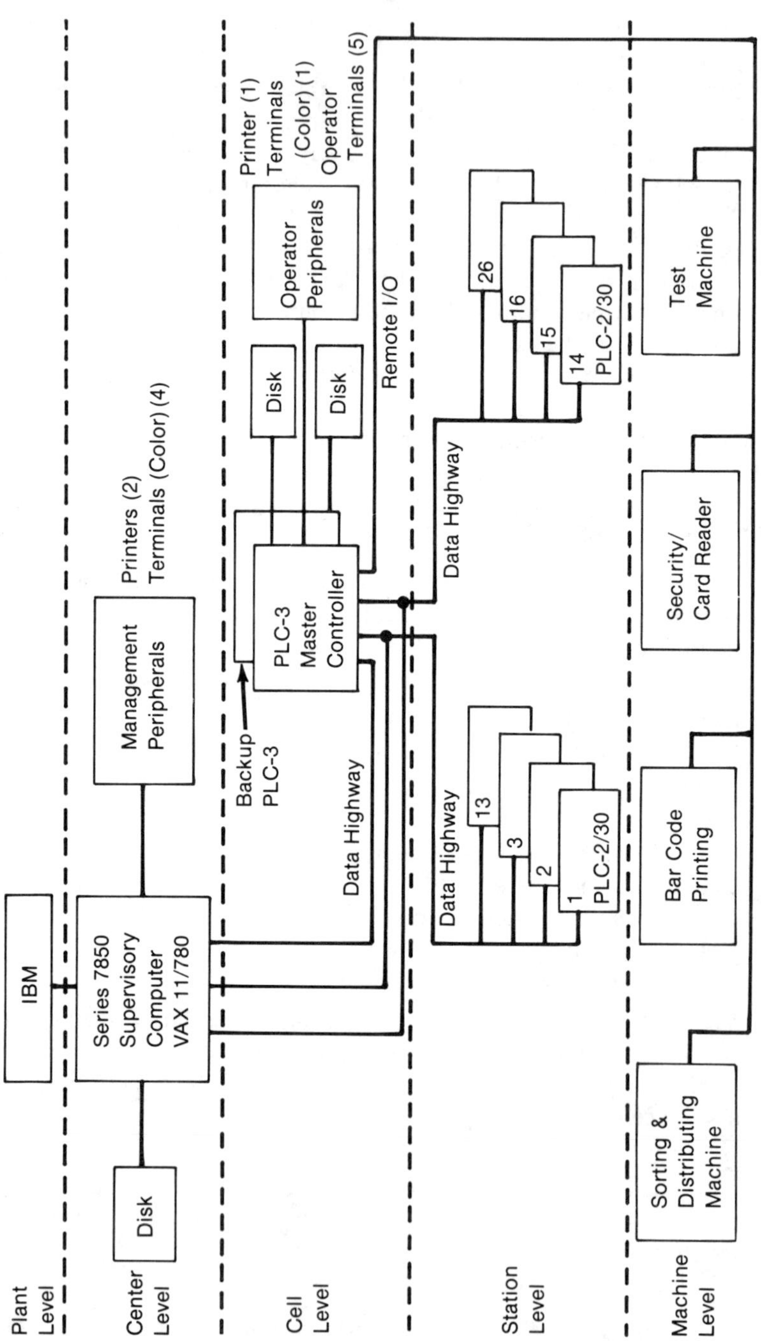

Figure 13-6. A Hierarchy of Control

(Courtesy of Allen-Bradley Company)

objectives. This is true, but — the same objection must have been made at one time about inventory control and about systems that coordinate purchasing, marketing, and production. Scheduling activities to accomplish manufacturing resource planning and computer integrated manufacturing are formidable tasks, but companies exist that offer software packages to accomplish them. If there are enough buyers to assure a profitable operation, the same kind of systems analysis will be put to work to produce software packages for hierarchical control.

Perhaps the greatest factor favoring a realization of information exchange between upper management and the processing areas is the present push toward distributed management, another name for the long dreamt of automated factory. The automobile industry, led by General Motors Corporation, feels that it must have this sort of organization in order to meet competition and is giving it the same intensive support that resulted in overwhelming weapons production in World War II. There are a number of roadblocks that are being recognized, and it is to be hoped that the efforts going into standardization of communication and development of software will reduce potential problems to produce methods for plantwide control that can be taken advantage of by the process industries.

Among the problems being investigated are the following:

(1) The automobile industry functions in the same manner as a batch-type process plant. The requirements for control of a batch process are not the same as for a continuous process, and so methods suitable for one may not be suitable for both.

(2) There is not yet agreement on the best type of communication for the automated factory. The choices of protocol and the designs of the data highway differ depending on whether information is collected from specific remote locations and processed at a central location, or whether a global data base concept is used, where a unit simply asks for information and the system goes and gets it.

(3) A lot of personnel level compromises will have to be made. There is the problem of line managers coming under close scrutiny by upper management who will have access to far more information than previously. There is the manager's fear of being replaced by a computer, or of losing his usefulness, his ladder to promotion. There is the problem of security — when corporate management has its finger on the keyboard, who will be allowed to change set points? Who is going to be responsible for the system? Will management become more control conscious, or will control and technical people become more management conscious?

The concept of total automation is being talked about for the process industry, for systems of PLC's, and for producers of machinery like General Motors and the John Deere Company. The important thing is that developments are taking place. It appears that the French expression "Plus ça change, plus c'est la même chose" — the more things change, the more they stay the same — may not be true in the context of industrial process control. Things are certainly changing, but they are just as certainly not going to stay the same.

REFERENCES

1. Urenon, Paavo and Theodore J. Williams, *Hierarchical Computer Control in the Pulp and Paper Industry*, Report No. 111, Purdue Laboratory for Applied Industrial Control. (1984)
2. Pait, Richard, and Kanai L. Ghosh, "Alabama River Pulp Co. Automates Recovery Boiler, Causticizing Areas," *Pulp and Paper*, April, 1985.
3. Cho, Chun H.,*Computer-Based Energy Management Systems*, Orlando, Florida: Academic Press (1984).
4. Dudzinski, Mark, "Digital Motor Drives Keep Paper Machines Humming," *I and C*, August, 1985, pp. 32–34.

5. Day, Garry, "What Can Dataloggers Do?" *Instruments and Control Systems*, September, 1979, pp. 41–44.

6. Ryan, Frank, Ronald Schabel, and Richard Radtkey, "Factor Equipment Capabilities into Load Management," *Power*, June 1985, pp. 91–93.

7. Thor, Merrill G., M. Robert Skrokov, and Marvin D. Weiss, "Energy Management in Process Plants through Distributed Control," *Instrumentation Technology*, March, 1979, pp. 30–32.

8. Leach, D. B., and E. F. Watson, "Supervisory Computer Control Cuts Steam Use 14 to 30 Percent," *I&CS*, June, 1984, pp. 63–65.

9. Hawley, Gary L., and George S. Carruba, "Computer Systems Provide Backup for Interconnected Grid," *Transmission and Distribution*, March, 1985.

10. Wells, Ronald D., "Computer Cuts Steam Cost by $1.3 Million," *InTech*, January, 1983, pp. 43–45.

11. Liptak, B. G., Editor, *Instrument Engineers' Handbook*, Section 7.5, pp. 713–723, Chilton Book Company (1985).

12. Blickley, George J., "Process Industries Take One Step at a Time," *Control Engineering*, June, 1985, pp. 83–84.

13. Laduzinsky, Alan J., "Factory Automation/U.S.A.: Just the Beginning," *Control Engineering*, June, 1985, pp. 88–89.

14. Morris, Henry M., "Is Anyone Interfacing Process Controls and Management's Computers?" *Control Engineering*, June, 1985, pp. 92–95.

APPENDICES

APPENDIX A — AMERICAN STANDARD CODE FOR INFORMATION EXCHANGE

ROW	BITS 4321	0	1	2	3	4	5	6	7	COLUMN 765 BITS ←
		000	001	010	011	100	101	110	111	
0	0000	NUL	DLE	SP	0	@	P	\	p	
1	0001	SOH	DC1	!	1	A	Q	a	q	
2	0010	STX	DC2	"	2	B	R	b	r	
3	0011	ETX	DC3	#	3	C	S	c	s	
4	0100	EOT	DC4	$	4	D	T	d	t	
5	0101	ENQ	NAK	%	5	E	U	e	u	
6	0110	ACK	SYN	&	6	F	V	f	v	
7	0111	BEL	ETB	'	7	G	W	g	w	
8	1000	BS	CAN	(	8	H	X	h	x	
9	1001	HT	EM	)	9	I	Y	i	y	
10	1010	LF	SUB	*	:	J	Z	j	z	
11	1011	VT	ESC	+	;	K		k	{	
12	1100	FF	FS	,	<	L	\	l	:	
13	1101	CR	GS	–	=	M		m	}	
14	1110	SO	RS	.	>	N		n	~	
15	1111	SI	US	/	?	O	_	o	DEL	

C, for example, is 1000011; K is 1001011.

The special symbols have the following meanings:

NUL	Null		DLE	Date Link Escape
SOH	Start of Heading		DC1	Device Control 1
STX	Start of Text		DC2	Device Control 2
ETX	End of Text		DC3	Device Control 3
EOT	End of Transmission		DC4	Device Control 4
ENQ	Enquiry		NAK	Negative Acknowledge
ACK	Acknowledge		SYN	Synchronous Idle
BEL	Bell (audible signal)		ETB	End of Tranmission Block
BS	Backspace		CAN	Cancel
HT	Horizontal Tabulation (punched card skip)		EM	End of Medium
LF	Line Feed		SUB	Substitute
VT	Vertical Tabulation		ESC	Escape
FF	Form Feed		FS	File Separator
CR	Carriage Return		GS	Group Separator
SO	Shift Off		RS	Record Separator
SI	Shift In		US	Unit Separator
SP	Space (blank)		DEL	Delete

APPENDIX B — ANSWERS TO EXERCISES

Unit 1

1-1.

The following are typical of relationships that can be computed.

a. In a flowing stream of fluid, **heat content**, as a function of mass flow, specific heat, and temperature.

b. For a distillation column, **internal reflux**, as a function of external reflux flow rate, top plate temperature, reflux temperature, and specific heat of the plate liquid.

c. **Mass flow of gas**, as a function of volume flow, absolute temperature, and absolute pressure.

d. **Stability** in a stirred tank reactor, as a function of reaction temperature, and the temperature difference between reactant and coolant.

e. **Fouling** of heat exchanger surfaces, as a function of fluid flow, and temperature difference between fluids.

f. **Power factor**, as a function of voltage.

g. **Efficiency** of a centrifugal compressor, as a function of pressure and temperature.

h. **Heat balances**, as a function of temperatures and mass flow rate.

i. **Optimized control action**, as a function of direction of error, and direction of correction.

1-2.

A process variable or a control output signal in analog instrumentation is represented by a range of 3–15 psi of air pressure or a range of 4–20 milliamperes of electrical current. Both electrical and pneumatic signals are subject to some degree of drift, nonlinearity, and hysteresis. A digital value, on the other hand, is a mathematical number that does not change and is explicitly repeatable. When the set point of a level controller, for example, is entered from a keyboard to have a value of 27.4, that value will be maintained by the instrument as long as the reference power supplies are functional. The value can be made to read 27.46 just as easily, something that would be impossible, or at least unreliable, by adjustment of a potentiometer or a dial indicator.

1-3.

The following must be known, so that they can be compared:
- Local labor and overhead costs for wire, fittings, and installation. The number of inputs and outputs.
- Local labor and overhead for panel wiring, and the number of terminations and cutouts.
- Costs for analog hardware; costs for digital hardware and software.
- Costs for control panel and control room.
  Intangibles:
  — Cost of down time, based on operating overhead, and estimated days lost per year.
  — An estimate of startup time and engineering time.
  — Cost of money, based on allowable payback period, and internal cost of money.

1-4.

Considerations include the condition of the present power source — how much interference from load changes and transients there is; the requirement for a clean environment for the operator interface — whether a new control room has to be built; operator acceptance of a new type of equipment; interfacing with the present system — assuring compatability of signals; how much of the present equipment and wiring to keep and how much to replace; timing and scheduling so that there is minimum lost process time; training; planning for expansion; and the choice of the equipment with which to make the replacement.

1-5.

Comparison between Leeds and Northrup Model C™ (electronic) and Electromax V™ (digital) controller.

	Analog	Digital
Input signal	0–4, 1–5 V	0–4, 1–5 V 0–10 to 0–100 mV Thermocouples, RTD's
Set point —		
Temp. stability	0.1%/30°C	6 microvolts/°C
Accuracy	0.5% (rms)	0.25% of input signal
Adjustment resolution	0.15%	1°F or °C, or 0.1% of range
Settings		
Proportional band	3 to 500	0.5 to 999.9
Reset, rpm	0.15 to 400	0.01 to 999.9
Rate	0 to 8 minutes	0.02 to 5 minutes
Lag	0 to 30 seconds	0 to 30 seconds
Environmental		
Operating temp.	0 to 60°C	0 to 60°C
Storage temp.	–30 to 80°C	–10 to 70°C
Relative humidity	90 to 40°C	90% at 40°C
Vibration	0.2 g at 0–60 Hz	0.5 g, 15 to 50 Hz
Shock	not listed	up to 10 g for 30 mseconds

Common mode rejection	not listed	at ref. conditions:
		low level input,
		   120 dB at 60 Hz
		high level input,
		   80 dB at 60 Hz

Normal mode rejection	not listed	50 dB at 60 Hz
RFI interference	not listed	less than 0.5% of set pt. or output span from a transmitter 1 meter away

Unit 2

2-1.

The addition of single bit numbers is shown below.

$$
\begin{array}{cccc}
0 & 0 & 1 & 1 \\
\underline{+0} & \underline{+1} & \underline{+0} & \underline{+1} \\
0 & 1 & 1 & 10
\end{array}
$$

As in decimal addition the 1 is carried to the next most significant digit.

Addition of positive and negative numbers is usually done using two's complement representation. Subtraction is accomplished by changing the sign of the subtrahend, and adding.

2-2.

Parity is used to check for errors in the transmission of binary data.

During transmission, electrical noise or failure of a component may cause a 1 bit to be transmitted or received as a 0, or vice versa. A parity generator adds the 1's and 0's of a word to be transmitted and puts a extra bit at the start of the word to make the sum positive (or negative, depending on the convention used). At the receiving end, a parity checker examines the word and generates its own parity bit, then compares it to the one received from the other end. If the parity bits do not agree, there has been an error in the transmission.

2-3.

A 5-bit word, to make 11010 for 26 letters.

2-4.

The least significant bit of a byte is the one representing the smallest binary division. By convention, this is the right-most bit in the word. The left-most bit of the byte is considered to be the most significant bit, even though it may contain a 1 or a 0 to establish a negative or a positive value.

2-5.

Tank agitator running	1
Pump No. 1 running	1
Pump No. 2 running	1
Pump No. 3 running	1
Bin weight OK	1
Tank empty	0
Drain valve closed	0
No connection	0

If the switches for the various input constraints produce signals for the necessary conditions as shown, the process can start when the input compares to 00011111, Hex 1F.

2-6.

The number 366 needs nine bits to contain the binary digits representing it, 101101110, so each day uses two bytes. May 19 is day 140, 8C in hexadecimal. AE20H + 8CH = AEACH.

Unit 3

3-1.

Bubble memory uses the presence or absence of "bubbles," magnetic fields (groups of atoms with parallel magnetic orientation) in a film of garnet, to represent 1's and 0's. The bubbles are propagated along paths determined by patterns deposited on an oxide layer covering the garnet, in response to a magnetic field generated by the currents in coils wrapped around the oxide layer.

The magnetic field is permanent, so no power supply is needed to maintain stored information. Bubble memory is solid state; it is not subject to contamination from dirt because the garnet film is encapsulated and is totally enclosed. It is not vulnerable to shock or vibration. It will operate at temperatures up to 70 degrees centigrade. All of these characteristics are beyond the limits of conventional disk drive memory storage.

Bubble memory has a very large storage capacity, larger than CMOS board-mounted chips. A single chip of complementary high density metallic oxide semiconductor (CHMOS) dynamic RAM with 256K bytes of memory on a single chip has been developed, but bubble memory can be several times larger.

More and more applications of bubble memory are being reported. Bubble memory has been used where rigorous environments make the mechanical motion of disk drives and tape systems difficult to maintain. Use of bubble memory reduces equipment size where very large memory storage capacity is required. Its use improves the speed of operation of a program because it is much faster to transfer data from an internal memory storage than from disk storage. Its principal limitations are high cost and a small number of suppliers.

3-2.

A compiler is a program that converts code written by a programmer in a high level language completely into machine language. An interpreter executes a program one line at a time, carrying out the commands of that line before going on to the next one. Assembly language code may be produced as an intermediate step. A compiled program runs much faster than one processed by an interpreter.

Source code is a program written by a programmer. It is converted by an interpreter or a compiler into machine language, which is stored in memory. This machine language code is called the object code.

3-3.

Physicists imagine an atom as being made up of a central nucleus of charged particles, with electrons rotating around it like planets in a solar system. There may be several layers of rotating electrons, but the outer layer determines the stability of the element and its capability to conduct electricity. If an atom has eight electrons in its outer layer, it is in equilibrium and is a poor conductor. If there is one extra electron, it becomes a conductor of negative current, which moves from one electron to the next. Or, if there is an electron missing from the rotating "shell," a positive current can be conducted, as the hole left by the missing electron is filled and emptied as the charge moves from one atom to the next.

Silicon has four electrons in its outer layer. Two silicon atoms share electrons, and the result is a stable, eight-atom outer shell. When silicon is doped with an atom that has one more or one less atom in its outer shell than the silicon has, conduction of positive or negative charges can take place.

3-4.

The figure below shows a two-phase clock signal used by the Motorola 6800™ microprocessor. The 8 events that are triggered by the cycling of the two signals fetch and decode a command, then get a byte of information from memory and load it into an accumulator. Phase 1 is the cycle for the address-bus clock and phase 2 regulates the data bus activity.

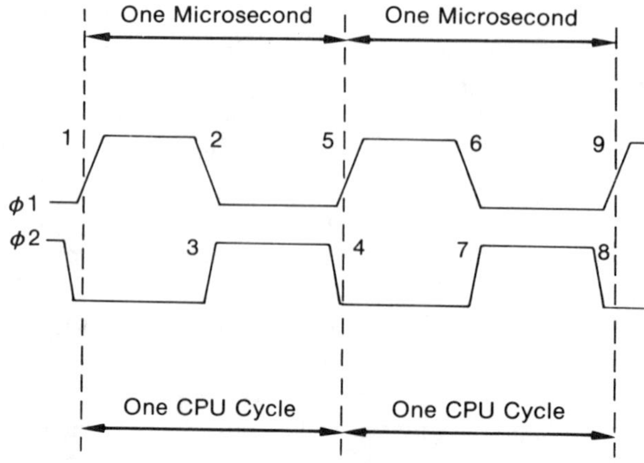

Figure A3-4. Timing for a Single Microprocessor Instruction

1. The address of the instruction is transferred from the program counter to the address bus. The necessary instruction codes for read/write are set to 1.

2. The program counter is incremented by 1 to the address of the next byte in memory.

3. The information in the designated memory address is put onto the data bus.

4. The data on the data bus is taken off the bus and decoded. From the bit arrangement, the microprocessor knows that the next byte it sees will be a number to be loaded, and not another instruction.

5. The address of the byte to be loaded is transferred from the program counter to the address bus.

6. The program counter is incremented by 1, to be ready for the next command.

7. The information to be loaded is transferred from its memory location to the data bus.

8. It is received by the microprocessor and placed into the designated register.

9. The beginning of the next cycle. The address of the next instruction is transferred from the program counter. Assuming a 1-MHz clock, all of this happens in about two microseconds.

3-5.

The decoder shown in Figure A3-5 monitors the 8 bits of an address bus, an input that is 1 when a command is to be recognized, and a read/write input that will be 1 when the decoder is to function. The decoder is designed to enable a device downstream when there is an address made up of any of the inputs A(8) to A(15).

When the read/write input is 1 and the validity input is 1, there will be no output from gate E if any of the inputs A(8) to A(15) are 1, but there will be an output if all of them are 0.

This could be used to differentiate between addresses above and below 00FFH in a 256K-byte memory.

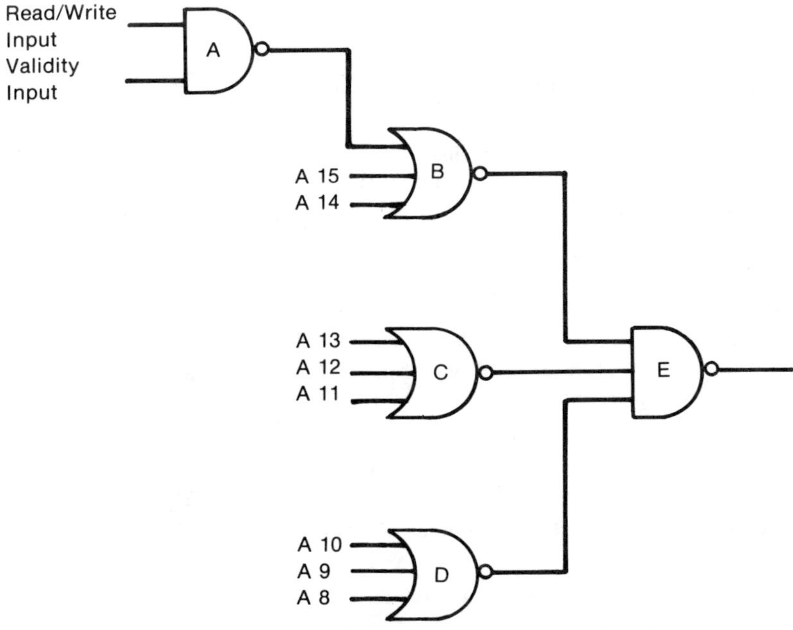

Figure A3-5. Address Decoder

Unit 4

4-1.

The ac power should come through a single breaker that is not at the same time supplying other loads. If the power distribution source also supplies power to large loads such as elevators and air conditioning units, there will be a constant change in supply level. If it supplies high frequency or intermittent loads such as arc welders and electric typewriters, the supply will contain electrical noise. Both conditions will contribute to poor performance of the digital control system.

Pay particular attention to grounding. The safety wire from equipment to ground should be connected directly to earth ground, not routed through a system used by other pieces of equipment. Faults in other equipment would introduce common mode voltages into the control system.

The best safety ground for a digital control system is a separate ground connection, such as a metal water pipe or a metal rod driven into the ground as close as possible to the location of the control equipment. If such a connection is made, the green safety wire should not be used for the digital equipment frame connection. Doing so could introduce two ground points into the system, making it susceptible to inductive pickup of electrical noise.

4-2.

The following considerations will enhance the design of a control room for digital equipment and CRT monitors.

a. Be sure that the ac power source is clean, free from transients and noise, and that it has ample capacity for present and future use.

b. Provide sufficient support areas adjacent to the operating area — room for spare parts, records, a service area, a separate office area.

c. Adequate lighting, preferably of low intensity and indirect to reduce glare.

d. To keep out curious visitors, isolate computers and peripherals, printers, or any other hands-off equipment, preferably in a glassed-in enclosure.

e. Provide easy access to the computer equipment for technical personnel.

f. Allow room for expansion.

g. Provide plenty of electrical outlets.

h. Provide suitable air conditioning and an adequate number of air changes per unit time.

i. Install wiring heavy enough to support all electrical loads.

j. Make sure the doors are large enough so that all equipment will pass through them.

k. Design panel and equipment mountings with shock and vibration in mind.

4-3.

The category of drawing sometimes called an instrumentation flow sheet is more properly called a P & ID (Process and Instrumentation Drawing). The Instrument Society of America has prepared a number of standards for symbols and documentation for such drawings.

ISA S5.1-1984, formerly ANSI Y32.20-1975, Instrumentation Symbols and Identification, applies to the entire process loop.

A standard, ISA S5.3-1982, has been prepared "to establish documentation for that class of instrumentation consisting of computers, programmable controllers, minicomputers, and microprocessor-based systems that have shared control, shared display, or other interface features."

For documenting interconnection of instrumentation, ISA S5.4-1981 has been prepared "to provide a method and practice for the preparation and use of instrument loop diagrams in the design, construction, checkout, startup, operation, maintenance, rearrangement, and reconstruction of instrument systems in industrial plants."

4-4.

A project engineer should be assigned who is familiar with the design, execution, specification, and engineering of digital control systems. He will, for any size job, need a designer familiar with commercial instrumentation to supervise the preparation of drawings and handle the details of the instrument, architectural, electrical, and piping disciplines. He will also need clerical help to prepare purchase orders, write in-plant specifications, keep records, and make schedules.

Larger jobs should include a staff of engineers so that the project engineer can concentrate on being a project manager, controlling finances and interfacing with the user. The staff may include instrumentation engineers, programmers, a quality control engineer, a building and utilities engineer, and an engineer representing the process department. There may be a counterpart to the project engineer assigned to field installation and startup.

4-5.

Common mode voltage is a differential in potential between ground at the sensor and ground at the digital control device connection. It is usually caused by circulating currents in the ground path separating the points.

Normal mode voltage is the voltage across the digital control device input terminals. If there is no error, this will be identical to the output from the sensor supplying the signal.

Unit 5

5-1.

Buses on printed circuit cards are parallel paths of foil. Each group of parallel strips is a bus specific to one function; data buses handle only data, address buses handle only addresses, command buses handle only commands. These are the three functions of buses that are usually required. If these same buses are built onto a backplane with the foil connected to pins of sockets, and if the edge connectors of the cards that plug into the sockets are connected to the same function bus as the pins they plug into, then any address bus (for example, on every card that plugs into the backplane) becomes common to the address buses of all the other cards that plug into the same backplane. The same will be true for the data and the command buses. By this means, a microprocessor on one card communicates with all the cards plugged into the backplane.

5-2.

Two informative articles comparing fiber optic cables and metal conductor cables are: "Fiber Optics or Electronic Cable?" by Ronald Stier and Ronald Ohlhaber, published in *I & CS, The Industrial and Process Control Magazine*, July, 1984, pp. 39–42; and "Comparing Costs of Fiber and Wire Local Data Networks," by Fred Eggleston, published in *InTech*, March, 1984, pp. 43–44.

Figures that follow are taken from these sources.

Cost Element	Twisted Pair	Coax	Lo-Speed Fiber Link	Hi-Speed Fiber Link
Interface (ea)	$500	$1500	$200	$850
Interface installation (ea)	50	100	50	200
Channel	50/ft	75/ft	1/ft	1/ft
Cable installation ($ per ft)	1-10	3-10	1	1

Cost Element	Twisted Pair	Coax	Lo-Speed Fiber Link	Hi-Speed Fiber Link
Performance:				
Max. bit rate				
(baud)	9.6K	1M	100K	20M
Max. distance				
(miles)	3	2	0.6	1

Considerations for selection:

Data integrity — If the EMI (electromagnetic interference) level will threaten your data, fiber optics may be the way to go.

Bandwidth — If you are using all of a fiber optic cable's bandwidth (which few industrial applications do), fiber optics can be the most effective choice. If you are not using all the bandwidth and all the length-bandwidth product that is available, electronic cable *can* be less costly.

Installed cost — For relative short runs (50 to 1000 feet) coaxial cable, twisted pairs, or multiple-pair cable will generally be most economical. What's more, they won't require the transmitters and receivers needed to interface to fiber optic cable. For longer runs, or where EMI is a real problem, fiber optic cable may be worth the additional expense.

5-3.

When you send a business letter by mail, it goes in a stamped, addressed envelope made within size limits defined by the postal service. A business letter will probably contain terms that might mean nothing to a layman but are understandable to another person in the same line of work as the sender. If the letter is going to a foreign country, there must be a further translation, or the receiver must be fluent in the language in which it is written.

Exactly the same demands of format and protocol must be observed in communicating between digital control devices. They must speak the same dialect of binary, or there must be a translation. The message must have a size and format that allows it to be recognized as a message. There must be an address. There may be a coded message to send back to the destination to indicate receipt, like the form that is filled out to acknowledge receipt of registered mail.

In brief, the objective is not to send a message, it is to send information and to send it so that its meaning is completely clear.

5-4.

A thermocouple consists of two dissimilar metals bonded or fused together. The junction produces an EMF dependent on temperature. Extension leadwires bring this EMF to a uniform temperature plane on a relay switching module in the data acquisition unit's scanner. Here it is referenced to the temperature of the termination panel and compensated for temperature and for errors caused by the junction of the thermocouple wires to copper terminals. The signal may be filtered at this point.

Next it is digitized, converted to binary form. Because thermocouple EMF outputs are nonlinear with respect to temperature, it will be digitally linearized, then stored in a register.

When it is to be printed, the binary value in the register will be moved over a bus to a port. It will probably pass in a parallel format to the printer input buffer, where each character will act as a pointer to a file that contains commands for activating the printer head. Other commands will move the head to the next print position and will finally perform a line feed that will index the paper into position for the next transmission, and a carriage return will move the print head to the left margin of the paper.

5-5.

In ASCII transmission, besides the binary combinations that represent characters and numbers, there are binary combinations that represent message control. STX, for example, means "Start of Message". To prevent STX from being mistaken for part of a message, it is protected by preceding it with another control form DLE, called a Data Link Escape. The receiver interprets DLE STX as a control form, not part of the message.

But DLE might really be part of the message. It was used in text in the preceding paragraph and would have to be specially identified if the text were to be transmitted. This is done by repeating it when the message is formatted, if it truly is part of a message. The receiver, when it sees DLE, waits to see if there is an immediately following DLE. If there is, it throws away the first one and interprets the second as a message element.

5-6.

The three methods are called frequency shift keying, amplitude modulation, and phase modulation.

Frequency shift keying is a form of frequency modulation. A carrier frequency is changed to one value to represent a 1 and to another value to represent a 0.

Amplitude modulation sends the carrier at a constant frequency but changes the value of the amplitude in order to signal changes in state.

Phase modulation shifts the phase of a transmitted signal by a specific number of degrees for each change in the bit pattern.

Unit 6

6-1.

Before any choice is made, all the parameters that must be measured should be defined. Then the uses to which the data will be put should be listed.

The following information should be known about inputs — number, source, signal conditioning required, ranges, and scan rates.

For digital inputs, the logic levels and frequency ranges should be known.

Uses to which the data may be put, in addition to data reporting, include averaging, comparison to limits, mathematical analysis, statistical analysis, control, and alarm-based decisions.

List peripherals needed, for example, printer, plotter, tape record, and computer link.

When the signals and uses are thoroughly understood, the following rough guidelines can be used for making a choice:

(1) Use a programmable data logger if the functions required include collection of up to 500 points of data; the throughput rate is less than 25 readings per second; limit alarms are required; real-time analysis and decision making are not required; there is a requirement to operate as a satellite for a distributed system.

(2) Use a programmable data acquisition system for inputs numbering between 100 and 10,000 if: the throughput rate is less than or equal to 1000 readings per second; real-time analysis and decision making are required; there is a requirement for a hardware priority interrupt structure; there is a need for direct memory access; one or two control outputs are required.

6-3.

For low level signals in high noise environments, the guarded integrating voltage-to-frequency A/D converter gives good results.

For very high speed measurements, the successive-approximation A/D converter is preferred.

The dual-slope A/D converter gives good performance in laboratory environments and for temperature measurements.

6-4.

Data loggers can monitor machine faults, air quality, and noise levels. Plant operators can be alerted by alarms to potentially hazardous conditions, allowing corrective action or plant shutdowns to be initiated.

6-5.

Before designing for optimizing, it is necessary to collect large amounts of baseline data. For boiler operation, for example, the data is used to determine maximum efficiency, to calibrate boiler instrumentation, and to establish operating limits. The Steam Generator Code, ASME PTC 4.1, and the ASME PTC 19 Supplements use the following data for making efficiency tests:

Main steam flow
Desuperheater flow
Blowdown flow
Auxiliary steam flow
Feedwater flow

Main steam temperature
Desuperheater spray temperature
Boiler feedwater temperature

Superheater outlet pressure
Drum pressure
Feedwater pressure
Desuperheater spray pressure

Ambient air reference temperature
Air heater air inlet temperature
Air heater air outlet temperature
Air heater gas inlet temperature
Air heater gas outlet temperature

Oxygen entering the air heater
Oxygen leaving the air heater
Fuel flow
Fuel temperature
Atomizing steam pressure
Atomizing steam flow

Forced draft fan outlet pressure
Furnace pressure
Moisture in the air

Fuel analysis, percents
    Carbon
    Hydrogen
    Oxygen
    Nitrogen
    Sulfur
    Ash
    Moisture
Fuel heating value

Unit 7

7-1.

Dead time is a finite length of time before a process responds to a change in set point or process variable. Visualize how it feels to drive a car that has "slop" in the steering mechanism. That is dead time, and you can imagine how it will affect your ability to steer and control the direction of your car in a straight line.

Even digital control cannot overcome dead time. In fact, a sample and hold circuit used in an extremely fast loop could introduce dead time, to the detriment of the control. The best cure is to reduce the dead time as much as possible.

7-2.

A control loop may have to operate for only part of a batch cycle. While a batch is heating, for example, the measured temperature of its temperature control loop will be below set point, and no cooling will be necessary. There is an error between set point and measured variable, however, and reset will drive the output as far as it can, trying to correct the error, until it reaches a limit.

When the temperature of the batch has become high enough so that the error between set point and measured variable becomes zero and then begins to increase in the other direction, cooling is needed. The output will begin to move, but it has to start from its limit, where reset had driven it. A finite amount of time must elapse before valve action begins to correct the deviation from set point. A batch process that is generating heat can overshoot badly in this amount of time.

7-3.

A process may have more than one controlled variable. The loops may interact with one another; a change in the manipulated variable of one may produce changes in the controlled variable of others.

Decoupling is breaking the interaction between loops. One method is to feed forward a signal from one loop into the others, to nullify the interaction. Because loops usually have different gains, this can become complicated even if there are only two interacting loops. It is the sort of calculation for which a computer might have to be used to get a satisfactory solution.

7-4.

Floating control and error squared control are algorithms that can be accomplished easily by programming in digital equipment.

Floating control is a type of control where there is no corrective action taken until deviation from set point increases above or below set point to a preset limit. For example, a tank level control might have a set point of 50%, with normal inflow and outflow sufficient to keep it close to that point most of the time. If the level rose above 75%, proportional control would function to reduce the level. If the level dropped below 25%, proportional control would function to increase the level. Anywhere between the two values, there would be no control, and the level would "float" at whatever value its self-control determined.

Error squared control is a form of proportional control where output changes as a function of the square of the deviation from set point. There is very little control action when the measured variable is close to set point, but more and more control action is produced as it becomes farther and farther away from set point.

This type of control action is very beneficial for certain types of system response. pH control is an example of a control system that can benefit from it.

7-5.

A diaphragm-operated control valve operates by balancing the pressure of operating air on a diaphragm against the opposing force from a spring. Valves may either open or close as the air pressure to them increases, depending on the valve construction. If air pressure or control signal fails, the spring action may make them go all the way open or all the way closed. This is called fail-safe action, and the process characteristics will determine in which direction the valve should fail. Direct or reverse action configuration should be selected for the PID controller to agree with the fail-safe direction for the valve.

Unit 8

8-1.

Using a watchdog system, each remote unit can monitor itself to assure proper operation. One way to test input, output, and operation of a system is to continuously cycle an output on and off and monitor the input and output to verify that the system functions. A way to test both input and output circuitry is to program a resettable timer that will time out only if it is not reset by the input from an external relay. This, in turn, is cycled by an output from the PLC unit.

8-2.

Evaluate flexibility and expandability. If it is limited in I/O and memory capability and in the type of configurations that can be programmed, it may not be easily expandable.
Verify that it can be integrated with other PLC and control equipment.
Verify that it has diagnostic aides, LED's, fuses, and removable wiring arms.
Programming should be the ladder logic type. Small PLC's are most likely to be serviced by electricians who understand ladder diagrams and will relate to this kind of programming.

8-3.

Count up the number of analog devices for the job and determine the number of analog I/O points. Count up the discrete devices. Add 20%. Find out from the manufacturer the number of words per point.

8-4.

Many systems houses have come into being to provide an answer to this problem. An excellant reference for lists of systems houses is the publication *Programmable Controls —The User Magazine*, 25875 Jefferson, St. Clair Shores, MI 48081.

Unit 9

9-1.

Reasons for using more than one CRT:
(1) To have a backup in case one fails.
(2) To give the operator more than one dimension to the viewing of his process. One CRT will normally display a critical group continuously, while the other can be used to watch trends, alarm logs, or overviews. It can also be used for off-line configuring.
(3) Two screens help for testing and debugging, where parameters show up in more than one display. Observing changes to process conditions on one screen while watching the results of relay action on another saves a lot of time when a system is being checked out.
The disadvantage is cost. The operator station, with its CRT and system electronics package, is by far the most expensive portion of the system. The larger the system, the more justification there will be for two or even three CRT's. Sometimes a user will buy a separate operator station prior to startup for use as a training aid. It then becomes available as a spare

and can be combined with the central station installation after the training need has been satisfied.

9-2.

The input terminals will be part of a remotely located electronics module that has an interface to a data highway. This module will have a station number so that the central station can call for information from the station by number. There will be a data base in the module where all information, including the input information in question, will be stored. It will be stored in a file that is named, and there will be a key (tag name, function, or some sort of code) that will locate it in the file. When the central operator station programming calls for function XYZ from station AB, that specific piece of data will be located and sent over the data highway to buffer registers in the operator station RAM. From there it will be incorporated into the display information package.

9-3.

Locate the CRT screen where room lighting is indirect so that glare from the picture tube surface is minimized.

Keep the screen clean. Viewers are tempted to point to the elements of the picture, and fingerprints accumulate quickly.

The CRT is more often affected by ac supply load changes than the electronics of the distributed control system. If there is a possibility that power will increase or drop below the allowable operating limits specified by the supplier of the equipment, a separate voltage regulator should be installed in the supply line to the CRT.

The answer to Exercise 11-3 is also applicable to this exercise.

9-4.

Wiring termination cabinets, or marshalling cabinets as they are sometimes called, permit field wiring to be installed and terminated before the electronics equipment is delivered and installed. This facilitates installation scheduling. It also provides an opportunity for using plant electricians to complete the wiring to the instrument cabinets. Some engineers believe that plant electricians are more likely to be familiar with instrumentation wiring techniques than contract electricians, who normally install ac power and lighting.

9-5.

When two keyboards are required, one is used for operator activities, while the other will be used for configuration.

Unit 10

10-1.

Input to computers used for business or for making scientific calculations is entered from keyboards, punched cards, or tape. Unlike the computer used for process control, these computers are not time dependent. Their programs have a definite sequence of operation and when they are finished, they shut down. The process computer must continually scan, sampling real-time data. Its program never shuts down. It is time dependent and responsive to interrupts from external events; it must operate from a real-time clock. Other differences are:

(1) The process computer will have a man/machine interface with the process.

(2) Business programs use large volumes of data stored in files. Real-time program tasks exchange small amounts of critical data between memory and input/output ports over fast buses.

10-2.

Integers are whole numbers, without decimal points or fractions.

Floating point values contain one decimal point. (25 is an integer; 25.0 is a floating point value).

Logical values are true or not true (false).

A single memory word (some systems use two, or even four words) having a fixed number of bits is used to store a real value. If the number of digits in the real value does not fit into the number of bits in the memory word, the value will be truncated, and errors can be introduced into computations using the value. Double precision allocates twice the normal number of consecutive memory locations for storage of real values, and both sets of locations are used together in making computations. This improves accuracy.

10-3.

Interrupts are signals to a computer to stop what it is doing, and to do something else. They may be external (for example, a contact closure) or internal. Interrupt service routines are programs pointed to as the result of an interrupt and brought into activity on a priority basis. An event causing an internal interrupt could be the pressing of a letter key on a keyboard. When the event occurs, the executive program checks to see whether any other higher priority interrupt is being serviced. If not, it stops all processing, saves all the important information from its microprocessor registers into a section of RAM called a stack, looks up in a table the instructions for the interrupt procedures called for by the specific event, carries them out, then restores the vital information from the stack and continues on as before.

10-4.

There are four categories of information to be considered when selecting a computer. They include computer hardware, computer software, the process interface, and management considerations.

Parenthetically, it is assumed that there is, available to the user, a programmer or analyst who understands the significance of the programming criteria, knows how to generate an operating system, and can write application programs. This kind of expertise is necessary to make some of the decisions.

Computer hardware considerations

Maximum memory size — How much will have to be stored on disk, and how much can be resident in RAM? Does this satisfy computing speed requirements?

Memory increments — How much can be added. and in what increments? 64K is a common size for memory increments.

What sort of memory protection devices are there?

What addressing modes are available?

Will the proposed unit have — and do you need —

    Floating point hardware

    Direct memory access

    External interrupt capability

    Automatic restart

How are diagnostics displayed? Are faults coded or clearly spelled out?

Computer Software

What languages are offered?

Do you need to be able to do on-line programming?

What sort of file manager capabilities are there? Can they support random and sequential files, with Create, Copy, Delete, and Append capabilities?

Is multi-tasking a requirement?

Process Interface Considerations

Does the machine have enough terminations for the number of inputs and outputs in the system to which it will be applied?

What is the analog-to-digital converter accuracy? 12 bits is common; 13 or 14 bit accuracies are available.

Do you need some analog outputs for recording, or for driving set points?

Will the machine interface to discrete inputs and outputs; pulse inputs and outputs?

Management Considerations

Reliability — How many successful installations, for applications similar to that being planned, have been operated using the machine?

Is the supplier well established?

How accessible are spare parts and service?

Should service be done on an in-house basis, an on-call basis, or should there be a service contract?

10-5.

A mathematical model is the mathematical abstraction of a real process.

A dynamic model can be used to continuously update set points and tuning constants of the loops controlling a process. Other uses include:

(1) The development of control algorithms

(2) Optimizing

(3) The determination of sensor locations giving best response to process variable changes

(4) The determination of appropriate sampling frequencies

(5) The selection of control (tuning) parameters for startup

Unit 11

11-1.

Computer systems can be programmed so that after the executive program has been activated (booted, in the parlance of the hobbyists), it cannot be used by an individual until a password has been entered from the keyboard. This also provides a means for keeping a record of the persons using the computer and the amount of time each person spends. The time may be charged as a function of departmental accounting.

11-2.

Software schemes that depend on encryption of the program and hardware techniques that operate through special-purpose plugs without which the software cannot function have been proposed to prevent copying. Special formatting of disks to alter the normal pattern of track and sectors is a method used frequently by programmers with a knowledge of the operating system and the internal commands that access the disk memory operation.

Numerous software programs can be purchased to use for protecting programs from copying. One of the more successful uses a "fingerprint" technique that permits copying only by the machine that implanted the fingerprint in the first place.

Of course, other programs can be purchased for cracking protection codes. There are those who maintain that it is impossible to develop a system that cannot be penetrated by someone else.

11-3.

CRT electronics are fan-cooled, with air drawn in through filters, to keep out room dust and dirt. The filters, if they are doing their job, will plug up. A routine filter cleaning operation is recommended as preventive maintenance.

11-4.

Every personal computer operating system includes some form of BASIC interpreter. Compilers can be purchased for BASIC, FORTRAN, Pascal, Forth™, C, and LISP™ for use on the two most popular makes of personal computer and many of the compatible models. Assemblers and macro-assemblers can be purchased for use by those programmers who work in assembly language.

For the user who does not want to learn to program, three types of commercially available software will provide the capabilities for almost any type of application: programs for word processing, spread sheets, and data base management.

Some software (for example, Lotus 1,2,3™ and Symphony™) combines all three in one package. Popular word processing programs are Wordstar™ and Word Perfect™. Most business applications can be handled by spreadsheet programs like Visicalc™ and Super-calc™. For users with file management needs, a program like dBase II will be useful.

11-5.

Major differences between printers for personal computers are:
The line length
The method of forming printed characters
The speed of printing
There are many other technical differences but these are the ones a user sees and the ones on which a selection will be based.

Line length will be either 80 or 130 characters. For letter size output, 80 characters fits an 8½ by 11 inch page. For data reports, 130 characters lines are practical.

The method of forming characters will be either by impact printing head — daisy wheel, ball element, or belt are common types — or by dot matrix head — wire dot, ink jet, or thermal or electrostatic charge. Impact printing produces the best quality print. Dot matrix, a halftone type formation, is good for anything except business letter quality work so long as the ink ribbon is replaced at regular intervals. Thermal and electrostatic heads require the use of special paper, and the characters are somewhat less distinct.

Speed of printing is a function of the type of printing head. Impact printing is much slower than dot matrix printing.

Speed is also a function of the amount of work the printer will have to do. 150 lines a minute is considered slow, 3000 lines per minute is considered fast, although there are laser printers that will put out 20,000 lines of printed material every minute. If many long reports have to be printed, speed is important. Otherwise, 150 or 300 lines per minute printing will satisfy most industrial requirements.

SUBJECT
INDEX